ON BEING A TENANT FARMER
A LAYMAN'S GUIDE TO THE LANDLORD AND TENANT SYSTEM

By the same author

Intensive Sheep Management

ON BEING A TENANT FARMER

A Layman's Guide to the Landlord and Tenant System

HENRY FELL

Farming Press
Ipswich

Published 1988

Copyright © Henry Fell 1988

British Library Cataloguing in Publication Data

Fell, Henry R. *1929–*
On being a tenant farmer:
a guide to the landlord and tenant system.
1. England. Agricultural industries. Farms. Tenancies. Law
I. Title

344.2064'348

346. 0434/

ISBN 0-85236-181/5

Cover design by Mark Beesley
Phototypeset by Galleon Photosetting, Ipswich
Printed in Great Britain by
Butler & Tanner Ltd, Frome and London

Contents

Acknowledgements

The author's name appears on the title page and, if credit is due, to him it is given. Any author knows full well, however, that he is but one among many who have made the work what it is. To all those who have so willingly, over the years, helped, inspired and supported me, I say a very sincere word of thanks.

In particular, I want to pay tribute to those very many people in the Tenant Farmers Association with whom it was such a privilege to work. My period as Chairman of the TFA gave me the opportunity to meet tenant farmers in every corner of England. Their generosity and hospitality were something that I shall never forget. A determination to repay that debt of gratitude was the main inspiration for writing this book.

To mention any by name is to take the grave risk of doing an injustice to many far too numerous to include. There are, however, two for whom I do not hesitate to take that risk. John Norris, who has contributed the foreword, was President of the CLA for much of my time with the TFA. We became friends for the good reason that we shared a common concern for, and love of, the countryside and its people. Eleanor Pinfold, Solicitor and Partner in the firm of Lawrence Graham, undertook the arduous job of reading the proofs and correcting any legal mistakes. To both of them, a very special thank-you.

Finally, this acknowledgement cannot be complete without recording my gratitude to the landlord and tenant system without which I could not have started farming.

HENRY FELL

This book is dedicated to my four sons, and their sons, that they 'may live as if they were going to die tomorrow, but farm as if they were going to live for ever.'

Foreword

The landlord/tenant system we have today is the result of experience and fine tuning over the last 200 years. It has worked because, by and large, the needs of both parties are recognised. It withstood boom and slump in the last century and the first half of this. The real problems began when politicians, looking for short-term advantage, began to meddle. Penal taxation used for experiments in 'social engineering' together with the folly of the 1976 Act nearly sounded the death knell of the system. If only politicians could look further than the next election.

Now as we enter more difficult times the attraction to landowning families—and institutions—of in-hand farming is diminishing. The surge in the use of alternative arrangements is witness to this. Many of these arrangements are but a half-way house to the traditional tenancy. They are evidence that landowners are keen to let but are unwilling to surrender vacant possession for fear of the consequences if they put all at risk again. Landowners should not be placed in this dilemma.

Henry Fell in this book ably demonstrates the advantages to both parties of the traditional system and the way it brings stability to the whole rural scene. The tenanted sector still has a vital part to play. It is the most certain route for new blood to enter the industry and provides the logical division of long-term and medium-term risk capital.

I endorse the note of hope on which Henry Fell ends this book. I add the further hope that improvements to the 1984 Act now being publicly discussed, together with much-needed reform of taxation, will encourage new lettings. But perhaps both CLA and TFA know that what is most needed is commitment to the system by politicians of all parties. After all, you cannot have tenants without any landlords.

J. H. M. NORRIS CBE

It should be noted that this book is written in the context of English law. Scottish law is different and those who live and farm in Scotland should be aware of this. The differences may only be matters of detail; but as I so often stress, attention to the detail is of critical importance.

CHAPTER 1

Introduction

This book is not written for the professional land agent, surveyor or valuer. They are very well catered for, both by training and by a score of learned books covering all aspects of the landlord and tenant system in all its complexities. Rather it is written for the ordinary tenant farmer whose skill, training and experience all lie in his farming. He cannot be expected to be master of two distinctly different professions. What it attempts to do is to provide the background knowledge, in layman's language, which will enable him to know when he needs professional help and how to set about getting it. At many points in this book, I shall stress the need for such help. I do not seek to supplant the valuer, very much the reverse; knowledge that a problem exists and can be solved leads to the need for skilled and sympathetic professional help.

I first took the tenancy at Worlaby in October 1959, at a time when it had been bought by one of the Scunthorpe steel companies. It is interesting to reflect, all these years later, that it was then a farm that no one wanted. Very wet, badly drained and virtually out of cultivation. But the landowner was prepared to share the cost of reclamation; I reckoned that it had potential; I wanted to farm: and the rent was only £2.50 per acre!

Now, nearly thirty years later, the farm is certainly very different. It is productive, though far from being the best farm in the world. It has given me a very satisfying life, and has enabled me to live in reasonable comfort and bring up a family of four boys in sane and healthy surroundings. And, not surprisingly, the rent has gone up!

The fact that I could get started in farming was due to the landlord and tenant system. There was no way that I could have got enough money together to buy a farm. It was difficult enough, goodness knows, to scrape together enough to stock it and keep going for the first two expensive years. And I had no ambition to get myself trapped on a smallholding from which I could never expand. For my father was an engineer, and it has never been easy for the son of someone outside farming to get in.

So a tenancy it was, and I have therefore good reason to be grateful for that system of land tenure which we have built up in this country. A system which, in many ways, is unique to Britain, and which is the envy of farmers and landowners in other countries of the so-called developed world. A partnership between owner and occupier, to the benefit of each. The owner providing that large sum of capital invested in the land at low rates of return which reflect not only its long-term investment value, but also the low risk in such a stable commodity. The tenant providing the risk capital to work the farm, the labour and the skill; leaving the owner free to enjoy the rather more leisured benefits of his property. But, above all, the greatest benefit of the landlord and tenant system has been the mobility which it has made possible in farming. It has enabled fresh blood to come in which would otherwise have had no chance. It has made it easier for those on a smaller farm to move up to a bigger one.

If all that sounds too good to be true, then I have to admit that that is certainly the case. Not all partnerships, any more than all marriages, are free from problems, even disasters. We are but human after all! Clearly there has been something amiss, for this century has seen a steep decline in the proportion of land in England held in tenancy. In 1900 around about 90 per cent was tenanted, whereas by 1980 that had fallen to approximately 33 per cent. And not only had that dramatic change towards owner occupation taken place, but all the evidence was that the fall was continuing at an accelerating rate. There were no farms on the market to let, and clearly there were very good reasons why landowners, when faced with a choice, would do almost anything rather than let to a tenant on a standard tenancy.

The reasons for this I shall discuss in a later chapter. But whatever the cause for this decline, here we were faced with the loss of a proven system of land tenure for which many of us had considerable cause to be grateful. There were persistent calls for legislation, and indeed it was widely recognised that much of the existing law was ill conceived. Naturally, not everyone had the same ideas on what the best solution was, and there was a risk that partisan interest would have too great an influence on the debate. What was agreed was that this represented the last chance to restore the landlord and tenant system to the position where it was a sensible and viable option for a landowner faced with the decision of what to do when a farm became vacant.

The Tenant Farmers Association was born amidst all this debate and public agonising. It was born quite simply because there was a growing feeling that the tenant's case was insufficiently well represented, even on occasions going by default. More than anything, if we tenants were not prepared to put our backs into it, we would have no cause to complain if the eventual solution was not to our liking.

It is very far from easy to start a new national association for whatever sector of interest, least of all in a scattered and highly individualistic industry such as farming. That it got off the ground and within five years became established and recognised as a responsible and influential body representing tenants is a very considerable achievement. Much of the credit is due to the energy and foresight of men like Dick Whittle and Stephen Hart, who put their heart and soul, to say nothing of a great deal of time and mileage, into its formative years. It surely proves that the need was there, and was not being met by any other organisation.

I joined the TFA in November 1981, a month after its initial launch, and I had the great privilege of succeeding Dick Whittle when he stood down as Chairman in February 1984. What I learnt during the three years that I held that office has led me to believe that this book should be written. I had thought that I knew most of what a tenant needed to know about landlord and tenant legislation. I know now just how ignorant I was, and how many mistakes I had made, many of them quite unnecessary. Perhaps more to the point, my travels around all the counties of England as Chairman of the TFA, meeting many hundreds of tenant farmers of all types and sizes, brought home to me the fact that I was not alone in my ignorance. I have been saddened and frustrated to see and hear of the consequences of mistakes that have been made and which could have been avoided. If this book helps to prevent even one such mistake, it will have been well worth the effort of writing.

CHAPTER 2

History of the Landlord and Tenant System

Like many of the great institutions which form the framework of life in a country with a long history, the landlord and tenant system in England and Wales had no one specific point of origin. Just as we have no written constitution, no Declaration of Rights, so for many centuries we had no countrywide written law governing the relationship between those who owned and those who farmed the land. There was no moment in our history when the landlord and tenant system was invented.

In medieval England, farming centred on the manor and society was strictly divided into different classes. The so-called manorial system of farming was based on three fields worked on a fixed rotation, each field subdivided into many strips cultivated by various peasant yeomen owing allegiance to their lord. This was subsistence farming at best; at worst, it was starvation. There were no reserves to carry over the winter, either in the form of money or, more important, as meat. It must have been a harsh life, but it endured for centuries and remained stable and adequate whilst the population in the country as a whole remained constant. There were, however, exceptions. The great areas of open heathland, for instance, in Gloucestershire and Lincolnshire, were 'ranched' with huge flocks of sheep. Wool, not meat, was the product and the great flockmasters, who were often yeomen of great entrepreneurial ability, traded with the Low Countries of Europe and became rich beyond their dreams. Evidence of their wealth and benefaction can be seen in the huge and lovely churches which grace these areas.

Change came largely because people were gradually moving to the towns. Food had to be bought and therefore produced and sold. Subsistence farming and the manorial system could no longer produce what was needed. And after the catastrophe of the Black Death when whole communities were wiped out and nearly one-half of the population died, numbers started to increase. What amounted to a population explosion occurred in the eighteenth and nineteenth centuries, and this, coupled with the increasing urbanisation of industrial Britain, led to inevitable and fundamental change in both farming and rural life. The period of 1740 to 1870 saw an agricultural revolution the like of which was not witnessed again until the post-war years following the great 1947 Agriculture Act.

As is so often the case, the need produced the men, the innovators and inventors, the husbandmen and the great livestock improvers, who led the way. Jethro Tull who invented the drill, the essential piece of equipment which made the Norfolk four-course rotation possible. The great landowners, Turnip Townshend and, after him, Coke of Norfolk, who led the way in demonstrating new ways of improving arable land. Robert Bakewell, the 'father' of livestock improvement. These men were certainly the best known, but there were many others who played their part in transforming the whole face of British farming in the course of a century. Literally, they laid the foundations of farming as we know it today.

It was no coincidence that the majority of these great improvers were large landowners. Possessed of great wealth, they alone had the capital to finance the cost of drainage, of marling and fertilising, of building and equipping. And of enclosure, that other essential in the transformation from subsistence to commercial agriculture. Undoubtedly, many of them benefited greatly. Between 1776 and 1816, Coke had so

6

improved his Holkham Estate that annual rental income increased from £2200 to £20,000, and yet, at the same time, made the fortunes of the tenants who paid these high rents.

It was at this point of huge investment and the consequent long-term improvement that the relationship between land-owner and working tenant farmer started to take on a new shape. Coke had the good sense to realise that his investment would not be matched by his tenants unless he granted them long-term security in order to enjoy the fruits of their labour. So long-term leases, usually for twenty-one years, became at least in the arable areas as important an element in improvement as the turnip crop, the drill and the better livestock.

There was, of course, no national pattern. But where individuals led the way, the counties eventually followed suit. By the 1820s, the so-called Lincolnshire 'custom' was established whereby an outgoing tenant received compensation for the value of the improvements he left behind. Here, I think, it is fair to say, Lincolnshire led the way, for this was surely one, if not the major, reason for the great strides made by Lincolnshire farmers throughout the first three-quarters of the nineteenth century. No tenant in his right mind was going to invest and reinvest in long-term improvements unless he could be sure of such compensation. The fact that the Lincolnshire custom was not backed by statute was of little importance. Landowners, even the largest, were resident, and the community had a common interest and a code of conduct which made any breach of trust highly unlikely.

Eventually, legislation came in the form of the Agricultural Holdings (England) Act of 1883 which upheld 'tenant right' for the first time, and gave statutory recognition nationally to the best of local practice. From then onwards custom based on practical experience and confidence in a working partnership was to be supplanted increasingly by legislation, a development which we will follow in the next chapter.

So legislation began to interfere, if that is the right word, in the operation of a system of land tenure which had developed to meet the increasing demands placed upon farming as it sought to supply the food for an industrialised and urban Britain. Those demands were met with tremendous success, and we can look back on the period in the middle of the last century as the 'Golden Age of British Farming'. All, however, collapsed in the last two decades of that century as cheap imports flooded in from the virgin lands of the Americas and Australasia. Farming went into a desperate depression accompanied by acute rural poverty which, with the exception of a brief respite during and after the First World War, was to last until the late 1930s.

The landlord and tenant system prospered in line with farming, and came to its zenith at the end of the last century, when probably over 90 per cent of the land in England was held on tenancy. The essential basis of partnership between wealthy and usually resident landowner and skilled working farmer provided the fabric on which increasing production and efficiency were created. And, let it be said, at the same time did much to create and enhance the beauty of rural Britain. Where today we need the Countryside Commission, the FWAG movement, Environmentally Sensitive Areas and Sites of Special Scientific Interest—all admirable in their way—previously all this was cared for by a class of people who saw it as their personal duty and pleasure to create beauty for future generations to enjoy. That they were able to do so was partly due to inherited wealth but also very much due to the landlord and tenant system. This system freed the landowner from the onerous day to day involvement of farming, and at the same time enabled him to take the long-term view of his vocational responsibilities, whilst his tenant, on the other hand, could devote his whole energies to the immediate worries of the day without the burden of the capital needed to buy the land.

So it was easy to understand the value of this sharing in times of prosperity. Could it stand up to the strains of acute economic distress? The fact that it did so, and moreover remarkably well,

8

should perhaps give us cause to pause and consider our attitude to land tenure as once more British farming faces a period of depression. It is difficult for us to comprehend the severity of the crisis in the 1880s. The fact that wheat prices fell from £25 a ton down to as low as £5, and over a very short period of time, should give us some idea of the scale of the disaster. Many, of course, could not make the transformation from prosperity to poverty. There was no shortage of tragedy, and the whole fabric of rural society suffered. Church and school, local tradesmen and craftsmen, all suffered from the obsession to buy at bargain prices in the world market. Land went derelict and farms fell vacant. But, it is fair to say that, in the main, the suffering was shared. Rents fell to virtually nothing. The resident landlord, albeit buffered by his wealth, was affected by the crisis along with his tenant. And he continued to shoulder his responsibilities for rural society. The worst effects of the depression were mitigated and the foundations remained intact, something for which two world wars gave us cause to be grateful.

Why then in the face of this history of solid achievement and success have we witnessed such a decline since 1900? Today little more than 30 per cent is tenanted, and surely such a fall must be taken to show that the landlord and tenant system no longer meets the needs of modern day agriculture.

That would be a dangerous and facile conclusion. The reasons for the decline are to be found in the nation's social policy rather than in any fundamental weakness in the system. Two basic mistakes can be identified. Firstly, the national obsession with 'social fairness', which has led successive Governments to penalise and harass the private landowner. A policy based upon jealousy which has done much to remove the traditional letting landowner from the rural scene, to be replaced by the owner-occupier or the City-based financial institution, both necessarily short term in their outlook. And secondly, really a consequence of the first, progressive legislation which has shifted the balance in the partnership in favour of the tenant. This culminated in the 1976 Agricultural (Miscellaneous Provisions) Act which granted two-generation succession to the vast majority of tenants. Small wonder that most landowners got the message that they no longer had a function as landlords! Happily, the 1984 (now 1986) Agricultural Holdings Act has begun to redress the balance, but there is a long way to go yet before confidence is restored. Happily too, society as a whole shows signs of realising that fairness, whatever that may mean, is not achieved by levelling everyone downwards.

We talk a great deal today about the need to conserve the countryside, and as a nation we are prepared to spend a lot of money and effort to achieve that aim. It is not unfair to say that we do not pay anything like sufficient attention to the central role of land tenure, in its widest sense, in the management of the countryside. Any study of the history of the last two centuries shows what an important contribution the landlord and tenant system has made. Surely it would be folly to let this go?

CHAPTER 3

The Development of the Law

I noted in Chapter 2 how the relationship between landowner and tenant came to be regulated by custom and subsequently by the individual tenancy agreement, which varied according to local circumstances. Desirable as that may appear when seen from a distance, modern economic conditions and levels of investment are such that that relationship must be closely and clearly defined by statute. That process really started in 1851, at a time of great agricultural prosperity, and successive acts followed. We can take our starting point most conveniently from the Second World War, and the 1947 and 1948 Acts which immediately followed it.

It is important to remember the emotional background to this legislation. We had just passed through six years of war during which on more than one occasion the nation had stood on the brink of defeat. The threat of invasion and the reality of aerial bombardment were met by our armed forces, whose efforts were nearly brought to nought by the threat of starvation. Our agriculture had been decimated by the depression which had started in the last century and was at its worst in the 1930s, so that our farming industry was in no shape at all to meet the sudden demands to feed our people as overseas supplies were cut off by the submarine blockade. The revival of farming production, and the herculean efforts that went into it, based on a labour force of women and old men, is one of the triumphs of the war. Our people were fed and we survived.

11

Although that story is not the subject of this book, it is relevant here because it brought forth the determination that never again would we run the risk of being starved into submission as a result of allowing our agriculture to fall into ruins. We can look back at that now with a certain sense of irony. Our problems today are all about regulating surpluses, a very far cry indeed from only forty years ago. Nevertheless, that emotional determination was the mainspring of the 1947 Agriculture Act, which aimed to put British farming on a sound and guaranteed footing so that the industry could develop and expand with confidence, irrespective of how world markets moved. That was the aim, and how wonderfully it succeeded.

The major part of the 1947 Act was all to do with prices and their annual review in February. But fundamental to achieving long-term confidence in the industry as a whole was the security given to tenants. Alongside price security, it was seen that occupation security was of equal importance if the necessary investments were to be made and justified. The tenanted sector was still in the majority. Whilst it had declined from the position at the turn of the century when it occupied 90 per cent of the land area, by the end of the war it was still over 60 per cent.

Something therefore had to be done about security of tenure. That something was to grant, retrospectively to all tenants, effective security for life subject only to certain conditions centred around the so-called Rules of Good Husbandry. Never before had a tenant been able to look at his farm and say to himself, 'This is mine to farm until I die provided that I farm it well, that I pay my rent, and that I fulfil the other obligations of my tenancy agreement.' Of course, it was enormously more complicated than that! But that was the gist of the matter, and fundamentally it was for good economic reasons, which were seen to override otherwise justifiable arguments against retrospective change in the law.

There was however a strong social element mixed up with the

12

economics. 1948 saw the beginnings, in agricultural legislation, of a policy based on the concept of 'fairness and equality'. The need was, therefore, to strengthen the position of the tenant and to disadvantage somewhat the landowner, who tended to be caricatured as both wealthy and privileged. And to be privileged in post-war Socialist Britain was to invite retribution.

This process of mixing up confused social thinking with necessary amendment to the law to meet changing economic and technological conditions was to continue right up to the 1976 Agricultural (Miscellaneous Provisions) Act. It was this act which swung the balance of advantage in favour of the tenant to the ultimate. The granting of three-generation security—subject, it must be said, to certain stringent conditions—had disastrous consequences. No landlord in his right mind was going to grant a tenancy which might well endure for as long as 100 years.

Thus the period from 1976 to 1984 was occupied by prolonged negotiations between the CLA and the NFU, representing the two parties. The Government made it plain that legislation would only follow when a broad consensus was achieved within the industry. Not surprisingly, the negotiations were far from straightforward, not only because a correct balance of advantage between landlord and tenant had to be found, but also because many other matters of detail had to be brought up to date.

In essence, however, the aims of the two organisations were simple. The CLA sought to get rid of three-generation security and return to one-life tenancies. The NFU, whilst agreeing that this was desirable, saw no reason why that should be conceded without something in return. That something was a change in the method of assessing rent at arbitration (and hence at negotiation). Since 1958 this had been based on an evaluation of 'open market' levels for similar holdings. Acute scarcity of land to rent had increasingly made this concept meaningless. There was no open market on which to base a

13

judgement, only a few isolated cases of 'tender' rents which had been pushed up to quite unreasonable levels in response to that scarcity.

Finally, in the spring of 1981, the CLA and the NFU concluded their negotiations and the so-called package saw the light of day. This deal having been struck between the two organisations, the Government agreed to prepare legislation, and a Bill, broadly based on the package, was published in October 1983. Naturally, this was not the end of the story by any manner of means. The debate continued both inside Parliament and out and, in consequence, important amendments were made to the Bill during its passage through both Houses. However, it eventually received the Royal Assent and passed onto the statute book on 12 July 1984. This now becomes an important date to remember for the practical reason that it marks the change point on, for instance, succession tenancies, amongst other matters of equal importance.

CHAPTER 4

The Agricultural Holdings Act 1986

The Agricultural Holdings Act 1986 is the basic Act of
Parliament which now controls the relationship between landlord and tenant. It brings together under one act legislation
which had been enacted over the previous forty years; that is to
say, it is a consolidation act which in itself contains virtually
nothing new. It covers the period from the 1947 and 1948 Acts
right through to the 1984 Agricultural Holdings Act, and it is
now the definitive guide, if that is the right word, for both the
professional and the layman. Moreover, it makes life much
easier for everyone in that there is no longer a necessity to hunt
backwards and forwards through a mass of cross-references. To
deduce from this, however, that interpretation is now simple
would be a gross exaggeration.

The remaining chapters of this book deal separately with all
the major elements of the act in some detail. Again, I must
stress that these chapters do not attempt to be an exhaustive
legal guide; rather they are aimed at highlighting important
matters which can be so easily missed and on which the tenant
may not even think of seeking advice.

In order to understand the 1986 Act, we must recall the
background of political negotiation and debate which led to the
Agricultural Holdings Act 1984, and which is outlined in
Chapter 3. For nothing that was achieved by the 1984 Act was
changed by transfer to 1986 other than a re-arrangement of
Schedule and Section numbers in the process of incorporating

15

all the elements of previous legislation which remained unchanged. This change in numbering can, however, cause confusion as, for example, with the notice served requiring the rent to be reviewed and an arbitrator appointed; previously this was known as a Section 8 Notice, and it has now become a Section 12 Notice. So beware of that pitfall!

Considering the ground that it covered, it is not at all surprising that the 1986 A.H. Act is both long and complicated, covering no less than 102 Sections. These are grouped together in logical order into seven different Parts, under titles that are largely self-explanatory.

Part I   Sections 1–5: Introduction and legal definitions
Part II   Sections 6–24: Provisions affecting a tenancy during its continuance
Part III Sections 25–33: Notices to quit
Part IV   Sections 34–59: Succession on death or retirement
Part V   Sections 60–78: Compensation at end of tenancy
Part VI   Sections 79–82: Market gardens and smallholdings
Part VII Sections 83–102: Miscellaneous

These Parts are followed by no less than 15 Schedules, which deal with matters of detail arising from principles stated in the Parts.

It sounds complicated. It is complicated. It is a job for the professional, but you have got to know when to shout for help. If you do not shout at all, no one can help you. If you do not shout soon enough, then not even the very best of solicitors or valuers can give the service you expect of them.

CHAPTER 5

The Tenancy Agreement or a Lease

As we have seen, the legal framework governing the landlord and tenant partnership is contained in the 1986 Agricultural Holdings Act.* But that relationship cannot rely solely on reference to an Act of Parliament which lays down the rules but does not deal with all the details of particular cases.

Also as we have seen, the detailed application varied according to different customs in different parts of the country. Common law and custom have now been largely over-ridden by legislation, but bits and pieces still remain. In my own county, for instance, the standard Lincolnshire Agreement was the undisputed basis for the contract of tenancy. And, moreover, it was accepted as the basis even if there was no written agreement between the parties.

We have, of course, moved on from there. Farming itself has become so much more complicated, and employs large sums of capital. Legislation, too, and not just agricultural legislation, has become more complex. So much so that it just is not sensible to rely on accepted custom or verbal agreement based upon trust and mutual confidence. Yet it is still the case that there are many instances where a written contract of tenancy does not exist. That has to be a mistake, if only because it can lead to confusion and subsequent bitterness. If you have not got

* Subsequently referred to as the 1986 A.H. Act to distinguish it from the 1986 Agriculture Act, cited as the 1986 A. Act, which deals with milk quotas and matters of related compensation.

an agreement, then get one. And if you cannot do so by mutual accord, then the law provides for either party to go to arbitration where the arbitrator is required to lay down an agreement. He is, of course, bound to conform to the ground rules as laid down by the 1986 A.H. Act; and, in doing so, he is instructed to reach an agreement which is both 'reasonable and just' as between landlord and tenant.

In essence, the agreement should cover the following headings:

(a) The names of the parties
(b) Description of the holding
(c) The duration of the tenancy
(d) The rent and details of when payable
(e) The incidence of rates including drainage rates
(f) Repairing obligations
(g) Insurance and re-instatement provisions by the landlord
(h) Insurance and re-instatement by the tenant
(i) A forfeiture clause
(j) Prohibition against assignment or sub-letting without consent

This is the skeleton and, needless to say, a great deal of detail needs to be filled in. And, of course, these heads of agreement form the basis for the more normal drawing-up of an agreement between the parties. But what sort of agreement? There is a great deal of loose talk, and very little understanding, about the difference between a tenancy agreement and a lease. Let me try to clear away some of the fog, for, as always, misunderstanding can lead to serious and potentially expensive mistakes.

Perhaps we should start by reminding ourselves that we are talking about agricultural land. That may seem to be a statement of the obvious, but a contract of tenancy can mean different things for agricultural and for non-agricultural land. This is becoming of increasing importance in these days of so-called diversification. Many possible uses of agricultural land are not in fact agricultural, in the legal sense, at all: a stud farm, for instance, caravan sites, even forestry. The use of land for any such non-agricultural use can take the tenancy of that land out of Agricultural Holdings legislation into the commercial law of landlord and tenant. So do watch it and seek advice before moving into any change of use. Having got that out of the way, and having our eyes firmly fixed on farming, we can pass on to a simple definition.

A 'contract of tenancy' means a letting of land for a term of years or from year to year. Thus the door seems to be open for a tenancy which can be terminated after a specified number of years. So let us take the definition a bit further.

A tenancy under the 1986 A.H. Act is an annual tenancy but with full security of tenure as defined in the act. The tenant has exclusive possession of the land. In the absence of any clauses to the contrary, where such are allowed, the contract will follow the provisions of the act.

A lease, on the other hand, is an agreement to let agricultural land for a specified term of years. However, at the end of the term of years, the lease reverts to an annual tenancy with full security of tenure. Thus a lease is not, in practice, a term tenancy as most people understand it, for it effectively continues to give the appropriate degree of security. I say appropriate because if it were a pre-July 1984 lease, the security would be based upon succession; if it were post-July 1984, it would be based upon lifetime security.

So if a lease reverts to a standard tenancy at the end of the term, why bother with a lease? A good question, and one which is ignored by many, thinking that it does not make any difference, but it very definitely does.

There are a number of areas where the terms of a lease may contract out of, or substantially alter, the legal framework of a standard annual tenancy. I must emphasise *may* contract out of, or alter, because of course the possible changes are not

19

necessarily obligatory. Some of the ways by which a lease may differ from a tenancy are:

1. The Term
The lease is really for a defined number of years; that is to say, the leaseholder or his successors in title have possession for that period of time; even if the tenant dies, the lease continues. The lease cannot be broken unless the tenant, by breaking the rules, gives cause for an incontestable notice to quit. This is in complete contrast to the standard tenancy where the death of the tenant triggers either the succession provisions or a notice to quit.

2. A Fixed Rent
The provisions for determining rent can be laid down in a lease on a very different basis from that found in Section 12 of the 1986 A.H. Act. There is the possibility of fixing the rent for the whole of the life of the lease. I cannot imagine any landowner or tenant in his right mind agreeing to that today, but it was certainly the case in the past. I have come across quite a few instances where a tenant was continuing to benefit from what now would seem to be a ridiculously low rent. One can imagine the frustration of a present-day landlord faced with a rent agreed in more stable days when inflation was not the norm.

3. Upwards Only
A fixed rent is becoming very rare, almost a matter of historical interest only. Of much greater importance is the current fashion for agreeing a rent which subsequently can only be re-negotiated upwards. This so-called 'upwards only' rent clause might well have been passed over in the 1970s and early 1980s as of no particular consequence, for those were the days when all in agriculture, from the Minister downwards, were convinced that everything was always bound to go up. Now, in the latter half of the 1980s, we are both sadder and wiser!

There are two possible variations on the upwards only theme: upwards only from the original base rent, i.e. the rent can fluctuate up or down so long as it remains above base, or upwards only from the most recent negotiation, that is to say, the best that can happen, from the tenant's point of view, is that the rent remains the same.

I have to say that I completely disapprove of such a system restricting the scope for negotiation on rent. The 1986 A.H. Act provides a workmanlike formula for negotiations, and it seems to me to be against the interest of both landlord and tenant to be put in a straitjacket which enables one party to take advantage of the other.

4. Open Market Rent

The 1986 A.H. Act was carefully drafted so as to ensure that a theoretical open market value was discarded in favour of a formula which included the productive and earning capacity (see Chapter 6). However, a lease may include a clause which specifically determines that open market should be the prime, or only, consideration. Here again, I cannot do other than disapprove.

5. Serving of Notices

Normally rent is reviewed and renegotiated once every three years. However, the lease may lay down any period of years or even that the rent should be index-linked. For a standard tenancy, or for a lease in the absence of a clause to the contrary, this process is triggered by serving a Section 12 Notice. But in a fixed-term lease it is perfectly possible for a clause to be included which enables the landlord to serve a notice requiring the rent to be reviewed, but which denies that right to the tenant. I can speak with some feeling on this example because it happened to me, as a lease I signed in 1978 contained such a clause. It never occurred to me that the day would come when I would be seeking a review with the hope,

21

and indeed the necessity, of getting a reduction in the rent. How innocent, and how ignorant, can one be!

This list of examples is by no means exhaustive. Indeed, it would not be possible, even given the time and the space, to produce such a list, for there are still many grey areas which have not been tested in the courts. That is far from surprising considering the scope and the extent of the 1986 A.H. Act. It will no doubt be a long time before everything is cut and dried—by which time no doubt we shall have more legislation, and thus fresh loopholes and grey areas!

The important question for the tenant is whether he should go for a tenancy agreement or a lease. The truth is that almost certainly he will not get the choice. Under current conditions in which the demand for farms exceeds the supply, the aspiring tenant will be presented with the terms of letting. But, at the very least, he should be fully aware of the implications of each and every clause and to do so, he must surely take skilled advice. And I would avoid an upwards-only rent clause like the plague. You may want to farm very badly, but there can be no justification for taking that kind of risk.

New tenancies and leases are one thing and, by the very nature of things, do not become available very often. For most of us, it is our current lease or tenancy agreement with which we are concerned. Can you honestly say that you have studied every clause and pondered its implications to you and your business in the light of current legislation and the present economic climate? If not, then you really must. And, further-more, you must do it regularly. As your family circumstances change—when the family grow up and you get older, when you alter your will—look closely at the document which has such an important controlling effect on the business which is your livelihood. And, moreover, get your solicitor to look at it with you.

CHAPTER 6

The Determination of Rent and Arbitration

The vast majority of rent negotiations are settled amicably by negotiation, and I would hope, and expect, that that would continue to be the case. Difficult economic conditions will almost certainly lead to an increase in the number of arbitrations, but it would be a sorry reflection on the responsible attitude of those concerned if that increase were to become a flood. Nevertheless, there is absolutely nothing to be ashamed, or frightened, of in going to arbitration. It is what I would call the last resort provided by the law for civilised men who cannot otherwise agree.

Depending upon the type of farm, rent is almost certainly one of the biggest items of fixed cost in the business. It is also one of the few over which the tenant has any hope of effecting a change, and it is the one which, theoretically at least, should move up or down in response to changes in farm profitability. It behoves the tenant to know exactly how the system works, and the 1986 A.H. Act gives detailed and, in some cases, precise guidance. It tells us how to 'trigger' the procedure and gives the appointed arbitrator instructions as to how he should arrive at his decision. It is important to note that, even though you have every reasonable expectation that you will settle by negotiation, it is desirable that you should follow the procedures laid down in the act. They provide, in any case, a sensible basis for that negotiation; and if you are not able to agree, then the job will be half done already.

Finally, one further general point, but so important: don't do it on your own. You may be the most skilful pig dealer in the business, but you can still be way out of your depth in this job. Use a skilled and trusted valuer. If you do not already know of one, then make use of one of the TFA's listed valuers. You can do so with confidence, knowing that they would not be on that list if they did not have the reputation of being skilled and fair-minded men. Moreover, I would give exactly the same advice to a landowner, for what we must be seeking is a fair and just solution and most definitely not for victory of one side over the other.

So what about the practicalities, the nuts and bolts of rent negotiation? Let me try to reduce it to the bare essentials.

THE FIRST STEPS

You can only formally renegotiate a rent once every three years. Naturally, that does not in any way prevent a landlord and his tenant agreeing mutually to some change, either temporary or permanent. It is not difficult to find examples where a sympathetic landlord has made some adjustment in the face of unusual, and quite unforeseen, circumstances. Indeed, my own landlords, and a City institution at that, agreed to delay an increase following severe flooding in the Ancholme Valley in 1981. A point to watch, however, in these instances of what one might justly call civilised behaviour is that a change in rent by agreement may well restart the three-year cycle from that point, despite the fact that that was not the intention of the parties.

Section 12 Notice
In the normal order of things, a rent agreed is a rent fixed for three years. That is not to say that it has to be renegotiated at the end of that term; merely, that it cannot be renegotiated

earlier failing mutual agreement. Even so, formal renegotiation does not occur automatically and as a matter of course except in commercial leases. It has to be triggered, and that trigger is a Section 12 Notice. This takes the form of a formal demand for a reference to arbitration under the act, and it must be served not less than one year and not more than two years before the term date. Note that it does not necessarily follow that arbitration will take place. Nothing in the serving of the notice precludes the two parties from reaching agreement.

You should note also the absolute necessity of the notice being received in time. The notice is not valid if it is late, and that means received, not sent. So send it in good time, and by recorded delivery, which means that the Post Office will obtain a receipt for you, or by registered post.

Contrary to what many people believe, either landlord or tenant can serve a Section 12 Notice (subject to any modifying clause in a fixed-term lease). Of course, during the years of both prosperity and inflation, rents were inevitably going up, and therefore logically it was the landlord who started the process going. But, in times of falling profit, the boot is on the other foot. If the landlord thinks that the rent may go down, there is obviously no incentive for him to serve the notice, so, in order to protect his interests, the tenant must do so. It is fair to say that there is a degree of reticence about doing this. Some tenants feel that it is not quite the 'done thing' to serve a notice on their landlords and that somehow it will sour the relationship. I also have to say that some landlords have encouraged this view. It is, however, totally illogical and, in any case, what was sauce for the goose is most certainly sauce for the gander. The fact is that in a business contract of the size of the average rent review, there is no room for such nonsense.

THE ESSENTIAL HEADINGS

One or other of them having set the process in motion, both landlord and tenant then have to prepare their cases. The 1986 A.H. Act lays down the basis which the arbitrator must follow. He must determine the rent at which the holding might be expected to be let by a 'prudent and willing landlord to a prudent and willing tenant'. In doing so, he must take account of four relevant factors:

1. The terms of the tenancy
2. The character and situation of the holding
3. The productive capacity of the holding and its related earning capacity
4. The current level of rents for comparable lettings

In doing so, the arbitrator is specifically instructed to disregard two important matters:

1. In so far as comparable holdings are concerned, he must disregard any element of scarcity, any element of premium payable or, finally, any effect on the rent due to so-called marriage value (see page 30).
2. He must disregard any increase in the rental value of the subject holding due to tenant's improvements or fixtures, as well as the grant element in any landlord's improvements.

No doubt all that looks rather daunting at first sight and, it must be said, it is still a matter for weighty debate amongst the lawyers, but if we take the four relevant factors step by step, then it only becomes a matter of organisation.

1. The Terms of the Tenancy
A statement of the obvious if you like! But wait a minute before you pass on. Some of the potential variables in the agreement

can have a significant effect on rent. Who, for instance, is responsible for repairs and insurance? If the agreement is 'fully repairing and insuring' (FRI)—in other words, the tenant is responsible for the lot—that could add as much as £10 to £15 per acre on to the rent as compared to a situation where the so-called Model Clauses operate, i.e. the landlord is responsible for all main structure and roofs.

Then, there is the question of drainage rates. Where they are payable, they are normally split into two fractions, owner's and occupier's. Is the tenant contracted to pay the owner's share? It can amount to a lot of money.

The shooting rights—does the landlord retain them, and is there a risk of game damage to crops? And so on. I think that I have said enough to illustrate the point that the tenancy agreement needs to be carefully analysed to cost out all the possible elements in what might be called 'hidden rent' or 'hidden benefits'.

2. The Character and Situation of the Holding

Again pretty obvious, you may say. Of course, but! And the buts can be important. The tenant will be fully aware of the pluses and minuses of the farm: the variability in soil type and how some fields, for some odd reason, always yield less than the others. As people are human, one side is always likely to claim that the farm is better than it is; the other that it is worse. The solution to this is to commission a full-scale soil survey by the Soil Survey of England and Wales. I can fully recommend this service, which gives a detailed picture of the farm's soils and their capacity and limitations. This completely independent and unbiased record of fact removes an area of dispute which is quite unnecessary. It is, without any doubt, much to the advantage of both parties to have this done and to share the cost, which is relatively modest in any case.

Then there are other elements apart from the soil. What plus value can you put on a farm being close to markets and a motorway network; or a minus value on the farm being very isolated; or close to the urban fringe with all the dangers of vandalism? All these and many other factors would go through your mind if you were seeking to rent the farm in the first place—and, if then, why not equally at the rent review?

3. The Productive Capacity and its Related Earning Capacity

Here the 1986 A.H. Act broke new ground. For the first time, the Arbitrator is required to take account of budgets and farm management information. Not as the sole criterion, please note, of the farm's ability to pay; but, certainly, as an equal element together with the other three, making up the four 'planks', if we can call them that, of the base on which the review is built. Neither more nor less important than what other people are prepared to pay on comparable farms—but equally important.

For the tenant, this is a major step forward. Previous legislation required the arbitrator to follow the open market in making his award. In a time of great shortage of farms to let, an open market hardly existed, and where farms were let by tender, rents were very high. In consequence, most arbitrators used their commonsense and awarded a figure which more nearly corresponded to their idea of the farm's economic capacity. Nevertheless this was an extremely unsatisfactory situation. Now that commonsense is enshrined in the law, the tenant can argue the farm's productive ability to pay with confidence. On the other hand, the rather naive hopes held by some that future rent negotiations would simply be a matter of arithmetic were always likely to be doomed to disappointment. The fact is that budgets now occupy one place in four, neither more nor less, and the market is still a most important feature.

That said, it is equally true that this part of the case must be prepared with great care and attention to detail. The productive capacity of the farm should be a matter of record, of fact and not make-believe. Figures of yield, both physical and financial, must be assembled, covering at least the previous three years and preferably five. And these figures will need to be supported by the actual evidence of sales notes, incomes, etc. In this way, a picture can be built up of what the farm has done and therefore can be expected to do. (See Appendix 3.)

Having done that, we can pass on to the related earning capacity. Here we have a dilemma, for we are not really concerned with the earning capacity of the last three years; it is the future, and the capacity to pay over the coming three years that is our interest. It is a matter of preparing budgets based on the farm's historical ability to produce multiplied by our expectation of what the produce will realise in the market place. Needless to say, there is plenty of room for disagreement there. Nevertheless, a pretty fair shot can be made at it, and, in any case, let us not forget that it is the valuer's particular expertise to size up this sort of projection.

In carrying out these calculations, one important fact must be remembered. The act speaks of the capacity of the farm in the hands of a 'competent' tenant. Not the actual tenant, nor a super-efficient tenant, and certainly not a below-average and inefficient tenant. The calculation must be built around what could be expected from the average competent farmer. So if, for some good reason, your yields are way above the average of the next-door farm, that can be discounted. Similarly, if you are a top-flight pedigree breeder earning large sums from the sale of bulls and frozen embryos, you are certainly not the hypothetical average competent tenant, and that part of above-average income can be set aside.

So a statistical picture is built up and by applying the appropriate levels of fixed and variable costs, we arrive at a margin before rent and before interest on borrowed money. This is the 'cake' which, at the end of the year, is available for division between the landowner (for rent), the bank (for interest), and the tenant (for his profit, out of which must come his living expenses, to say nothing of the build-up of a capital replacement fund). There are those who argue, very strongly, that bank interest should be included as a valid cost before the calculation of the cake. I can see much logic in this argument, but it is not one which is accepted by the valuing profession. Be that as it may, we have a cake for division, and here the Act lets us down. Nothing is said, no guidance is given, as to what is a fair and reasonable split of the cake. Should it be as in days gone by, one-third for the rent and two-thirds for the tenant? Or should it be, as seems to be the case today, a fifty-fifty split? There are even those who go further and claim that the tenant's share should be no more than 40 per cent, but that seems very close to greed to me.

So the arbitrator is left on his own to make his judgement. And perhaps that is after all the right way to do it. He is not a mechanical calculator; he is an experienced valuer who uses his knowledge and judgement. But, my word, it places a burden of responsibility on his shoulders.

All this assumes, however, that the case will go to arbitration. As I have noted earlier, the majority will undoubtedly be settled before that final point is reached. How then is a decision to be made as to how this cake should be cut? The answer of course is that, at this stage, we have entered the field of bargaining, of strength and skill of advocacy, and of bluff and counter-bluff.

4. Rents on Comparable Holdings

Here we come to what some regard as the real issue—what other people can and will pay, in other words the open market. But it is an open market with a difference, for we must not forget the important disregards. Any element of premium paid in a tender rent must be disregarded, as must that element attributable to scarcity. So too must the effect of 'marriage value', which is where the comparable farm is one of a joint holding where the rent payable is higher as a consequence of economy of scale. We must note that the farm must be comparable, or nearly so. If comparisons are to be made, then obviously only like must be compared with like. Otherwise a farm submitted as comparable can be very easily attacked during the hearing. Clearly, what at first sight seems a good idea—'The man three miles down the road can pay £50 per acre so why can't you?'—is not as simple in practice as it sounds. Nevertheless, each party will want to produce a suitable comparison if possible, and for the tenant it will greatly support his contentions about the productive capacity of the farm if he can show that the next-door farm really does no more than he does. Again factual, evidence-supported yields are impressive. Pub gossip is worse than useless!

THE NEGOTIATIONS

So we have reached the stage where the preparatory work has been done and the facts assembled. The twelve months between the last day for serving the Section 12 Notice and the term date may seem a long time. And indeed it is ample time provided that the ground work has been well prepared and that nothing is left until the last minute. When should negotiations start with the landlord's valuer? A matter of judgement certainly, and one which both parties will be considering. An element of psychological pressure will no doubt come in at this juncture. When profits are rising and farming is buoyant, the tenant will want to settle as soon as possible. But, when land values are falling and the outlook is gloomy, the tenant will prefer to leave it until the last minute, believing that the pressure building up will be to his advantage. But please do not wait literally to the last minute. It is better by far to have plenty of time to explore each other's minds, to try to estimate what the limits of negotiation really are as opposed to the first, and no doubt unrealistic, offer.

It would be surprising if settlement were to be reached at the first meeting; indeed, it would probably be unwise. A period for reflection and an evaluation of tactics is highly desirable. But the time will come when both parties have to see whether there is a suitable basis for a settlement. Although it is possible that they will have arrived at the same figure, it is far more likely that there will be a gap, and then the real thinking has to begin.

The tenant and his adviser will have to estimate their chances of doing substantially better by going to arbitration. I say substantially better because they must weigh up, not only their chances of success or failure, but also the costs that will be involved. There is little point in gaining only a minor improvement in the offer which is currently on the table if the costs are going to more than mop that up. It is a matter of fine judgement, and there is no nice easy formula to help you work it out. If your valuer is the experienced professional that he should be, he will be able to advise you. Remember, though, that the landowner and his advisers are in exactly the same position; they face the same dilemma even though they are almost certainly in a better position to carry the costs.

THE APPOINTMENT OF THE ARBITRATOR

We have got to the point where, unhappily, it has not been possible to reach agreement, so the procedures for actually going to arbitration must be started. The two parties have a choice: either they can mutually agree on the appointment of some third party to act as arbitrator, or failing that, one of the parties must make an application to the president of the Royal Institution of Chartered Surveyors.

If the former, the procedure is simple. A person is agreed, he is approached and he consents to serve or not, as the case may be. No costs are involved and there is no problem provided the arbitrator is appointed before the term date. That date is crucial.

If the latter, the correct formal procedure must be followed. A written application must be received by the president of the RICS before the term date, but no longer than four months beforehand. The application must be accompanied by payment of the current fee, which at the time of writing is £70. Note that the application is invalid either if it is late or if the fee is not paid. Note too that either party may make the application for the appointment. Normally, of course, it is the party which originally served the Section 12 Notice, which after all was the instigator. But circumstances can change, and change dramatically, in twelve months, and the boot may well be on the other foot. There is nothing to prevent one party riding on the back, as it were, of the other party's Section 12 Notice. And once that notice is served there is no going back unless both agree to call it a day.

In due course then, but not necessarily before, and almost certainly after, the term date, the president will notify each party of his choice as arbitrator. That choice will be made from a list of approved persons assembled by the Lord Chancellor. They are professional valuers from all parts of the country, and are recognised to be men of stature. There is talk, naturally enough, of arbitrators being known as either 'landlords' men' or 'tenants' men'. It is not surprising that, being human, their sympathies may lie in one direction or another; but they are appointed to a legal position, and there is little justification for either landlord or tenant to feel that he will get anything less than justice.

THE STATEMENT OF THE CASE

The arbitrator has been duly and correctly appointed and has notified the parties of it. They then have a maximum of 35 days in which to submit to him a statement of their case. This should be as concise as possible; nothing could be better designed to irritate the arbitrator than to have to plough through pages of fluffy verbiage. It should follow the format as laid down in the 1986 Act, and proceed logically from step to step. It may, or it may not, include material or correspondence which passed during the negotiations. Such material incidentally will have been marked by your adviser as being either 'without prejudice' or just left 'open'. The point of this is quite simply that the without-prejudice material cannot be used against you, or by you, at a later date.

At this point it is once more necessary to emphasise the importance of having a skilled valuer at your side. Few tenants have had any experience of writing a statement of case or going before an arbitrator. Not all valuers have either, for that matter, which underlines the absolute necessity of employing the right one.

THE ARBITRATION

The statements of case having been duly lodged with the arbitrator, he will nominate a date and place for the hearing. At this stage, each party will receive a copy of the other's statement, together with copies of any correspondence that passes between them and the arbitrator. Each side can then see and make judgements on the grounds on which the other is going to fight. A time indeed once more for reflection and a reassessment of strategy. It should be said that there is absolutely nothing to prevent a settlement being agreed at any time right up to the actual moment when the arbitration starts. Indeed, the mind is often concentrated most wonderfully by the imminence of proceedings starting, and it is by no means uncommon for agreement to be reached at 'five minutes to midnight'. Costs will of course be involved from the arbitrator for the time and trouble to which you have put him, but they will be relatively minor compared to the costs of the arbitration itself. No one should be too proud to agree at the last moment if that is the right thing to do.

Let us assume that agreement is not possible, and the

arbitration goes ahead. Let me give you an outline of what will happen.

Remember that although this is not a court of law, county court rules still apply, even though the hearing may take place in a farm house or the room at the local pub. Don't be misled by the local surroundings. Evidence will be given under oath, and each party is free to be represented by whom he chooses. At its simplest, the tenant could represent himself, although he would be very stupid to do so. At its most complicated, and most expensive, each party may choose to use a solicitor or a solicitor and a barrister in addition to his valuer.

Each side puts its case, is subject to cross-examination and, finally, has the opportunity to sum up. As in any court hearing, the power of advocacy is of the greatest importance. You may have a case which is full of solid fact, but if it is badly argued and if you stumble and falter under cross-examination, you will not make a good impression. That is not to say that you have to employ the most able, and thus most expensive, solicitor or barrister in the land. It does mean that you cannot afford to be represented by the inexperienced.

So the hearing comes to an end. Meanwhile, the arbitrator will want to visit the subject farm, as well as those submitted as comparables. You must then sit back and wait with as much patience as you can muster.

THE AWARD

Normally, the arbitrator will announce his award by sending it to both parties within 56 days of his appointment. However, this time limit may be extended by permission of the president of the RICS, and it is by no means unusual for this to happen.

When making his award, the arbitrator can be required by either party to give his detailed reasons for it, and it is now accepted practice that he should do so. The award is binding on both parties, and there is no appeal except for a challenge on a point of law. That apart, you are stuck with the decision, and there is nothing that you can do about it.

COSTS

The arbitrator has discretion as to the award of costs. If he believes that both sides have acted responsibly and that a genuine case of difficulty has been brought before him, then he will normally direct that each party should pay its own costs, and that his costs should be split equally between them. However, if he has reason to believe that one party has acted unreasonably, then he has the power to apportion the costs in any way he chooses. It is highly unlikely that 100 per cent of all costs would be charged to one party, but a 70/30 split is not unknown.

How much will it cost? I cannot possibly say—it is akin to asking how long is a piece of string. You must take the advice of your valuer, who will be in the best position to make an informed guess. It is very unlikely, however, that it will cost you less than £2,000, but equally unlikely that it will reach the £10,000 to £15,000 which are the subject of scare stories put about by some agents who ought to know better—unless of course you use valuers, accountants, solicitors and barristers, and the matter ends up on a point of law in the House of Lords.

Unfortunately, it has to be said that the cost is not proportionate to the size of the farm. It is really not much more expensive for a 1000 acre farm than for a smallholding. Thus the scales are weighted against the small farmer. An increase of £5 per acre on a farm of 100 acres is only £500—I say 'only' although it may be sufficient to make a painful difference to the profit—but measured against the arbitration costs it is little enough. £500 over three years is £1,500, which is no more than the costs if as much. A cruel dilemma borne by the small farmer, but much more easily resolved by his bigger neighbour.

CONCLUSION

The end of a long chapter which, try as I might, I have been unable to condense any more. As it is, I am conscious that I have only dealt with the bare bones; and aware too that each situation is individual and different. But it is equally true, as I said at the beginning, that there is probably no other subject which is of such importance to the tenant in the conduct of his relations with his landlord. Without any doubt at all, it is the subject that will come up the most regularly. It is essential that you get it right.

CHAPTER 7

Succession

THE BACKGROUND

As we have seen, tenants' security of occupation of their farms has increased over the years and following a whole series of successive acts of Parliament. Originally, a yearly tenancy was exactly what it said it was. Clearly, the lack of any sense of stability, the lack of any personal and emotional attachment to the farm, was in no way conducive to either good farming or investment in improvements. What might have been suitable for the medieval peasant farmer was quite unsuitable for his modern counterpart.

This move towards increased security of tenure came to its peak in 1976 when succession tenancies became law (see Chapter 3). The purpose of this legislation was to extend the lifetime security of the tenant beyond his death to cover the next two generations, so that, provided certain rules were followed, a farm could, in reality, be let to a certain family for a period of perhaps 100 years. The law had indeed come a long way from the days of the 1930s when a tenant could be turned out of his farm at no more than a year's notice and for no particular reason.

Was this legislation an essential piece of social justice in keeping with modern ideas of what was fair and just? It was certainly seen that way by many of those in the Labour Government of the day whose thinking was more idealistic than practical. There were others who saw it as a means of expressing their jealousy of the private landowner who they believed possessed privilege that was quite unacceptable. The 'exploitation of the downtrodden tenant by the wicked landlord' was seen as an evil on a par with that in the towns where an unsavoury case had come to light surrounding a Mr Rachman, whose name subsequently passed into the English language.

Such emotional judgements, however, are not in keeping with the clear-headed and practical thinking which is essential if the structure of the countryside and its major industry is to be both conserved and improved. This piece of legislation, giving what its authors imagined, no doubt quite sincerely, to be a substantial benefit to tenants, turned out in fact to be greatly to their disadvantage. And, moreover, very nearly sounded the death knell of the landlord and tenant system.

The reason for this was quite simple, as George Lillingstone, then President of the CLA, made clear in a paper to the Oxford Farming Conference. No landowner in his right mind would consider letting a farm when that meant losing the chance of getting vacant possession for possibly 100 years. And so it turned out that the supply of farms coming on the market to let dried up completely. Existing tenants might be secure and could see their sons and grandsons following them. But what of other sons; and, much to the point, what of able young men and women whose fathers were not farmers? The tenanted sector became frozen, and the mobility which was one of the great advantages of the landlord and tenant system was lost. And the move towards taking land in hand, or farming it with a faceless management company, accelerated.

All this soon became quite obvious. It led to the CLA/NFU agreement which eventually became the major part of the 1984 Act in an effort to undo the damage. This, however, was not the end of the story, for, whilst the 1976 Act which created three-generation tenancies was retrospective, the 1984 Act very properly was not. We therefore still have succession tenancies and the purpose of this chapter is to explain what the principles are.

THE PRINCIPLES OF SUCCESSION

The first thing to understand is that succession tenancies apply to the great majority of, but not all, tenancies created before 12 July 1984. And unless the law is changed retrospectively, which is highly unlikely, they will continue to apply until the last generation dies out. However, in addition to post-July 1984 tenancies for which there are no succession rights*, there are certain important exceptions:

1. Where a valid Notice to Quit has been served for a reason other than the death of the tenant
2. Under grazing licences and other such arrangements outside the provisions of the 1986 A.H. Act
3. Where two previous formal successions have taken place
4. Most County Council smallholdings

So an essential first step for any pre-July 1984 tenant is to confirm whether his tenancy has potential succession rights. If he has, what does he do next? In essence, he has to ensure that the following conditions are met:

1. That the potential applicant is *eligible*, and that is taken to mean:

- that he or she is a close relative, i.e. husband/wife, son/daughter, brother/sister, a natural or adopted child or a child who has been treated as a member of the family since childhood
- that he or she has obtained the greater part of his or her livelihood from the holding during five out of the seven years

---

* There are a few cases where a post-July 1984 tenancy can carry succession rights, but the vast majority do not.

41

immediately preceding the succession (note that this can include three years of full-time further education)
- and, finally, that he or she is not the occupier of another viable commercial unit

2. That the potential applicant is *suitable*, and that is taken to mean:

- that he or she has had suitable training and experience, and this usually means that the applicant has had some formal agricultural training in addition to being seen to be a competent farmer
- that he or she is in sound health
- that he or she is financially secure with access to sufficient capital to finance the business

These rules of so-called 'eligibility' and 'suitability' are fundamental, and even the briefest reading of them should make it obvious that succession is not automatic. In cases of dispute—and remember that it will be much to the landowner's advantage to break the succession, thus giving him vacant possession value—the right to succeed will be tested before the Agricultural Land Tribunal (see Appendix 2).

What should also be obvious is that definite and constructive planning for succession must not be put off; it must be done immediately, right from the beginning. Furthermore, these plans must be updated at regular intervals.

An Example

Clearly, succession bristles with complications of all sorts. It would be tedious in the extreme if, in this book, I tried to cover every possible option. Far better, I think, for me to try to illustrate the practical lessons to be learnt by taking a hypothetical family as an example. But only as an example, remembering that every situation is personal and requires specific advice.

A young tenant, in his early thirties, is married with three children, a boy and two girls. He will be looking forward to farming until he retires at some distant future date, when he hopes his son will follow in his footsteps. No need for him to bother his head about who will succeed him for, like most of us at that age, he is quite convinced that he is going to go on, if not forever, at least for a very long time. Sadly, disease and accident can intervene. In the same way that one hopes that he has taken out adequate life insurance, so we must insist that he plans for succession. Clearly, his son is too young and will be for many years, so will his widow be both eligible and suitable? The 1984 Act made it much easier for widows to succeed, but she still has to qualify. Has the will been written in such a way that she will have adequate finance? Does she earn any substantial income from outside the farm business? The answers to these questions must not be taken for granted. And supposing that for some good reason she could not qualify, is there a brother who could? That could, of course, be much more difficult for, unless he works on the farm, he risks failure on the earnings rule.

In any case, succession to a brother is potentially unsatisfactory. Apart from anything else, future succession would pass down his side of the family so that the object of preserving the farm for the widow's son is lost. It is far, far better for the succession to go to the widow, and this emphasises how esssential it is to ensure that a wife is qualified should her husband die before any sons or daughters are qualified.

Let us now move forward in years with our example. The children are now in their early twenties, but their father is only fifty-four. He does not want to retire yet and, in any case, what else could he do? The farm has been his abiding interest and will continue to be for many years. His son, now aged twenty-two, has been to college and, like his father, is mustard keen on farming. But he is also ambitious and frankly the farm is hardly big enough to contain both their enthusiasms, let alone provide both with an adequate income. Naturally, too, he wants to prove himself, to be his own man, and not to be dependent on Dad. At twenty-two, he can qualify for succession. Two years at college and three years working on the farm cover the minimum five years. He is generally recognised in the area as a competent and very promising young farmer. But he is chafing at the bit and needs to do his own thing. What about starting an agricultural contracting business? After all, the forage harvester and the big baler are not fully utilised at home. And that powerful new combine that he would so dearly like to buy would be a proposition if an extra 400 acres could be found locally.

Wait a minute! What about the earnings rule? Isn't our son just about to make himself ineligible? Wouldn't it be better if two separate companies were formed, one for the farming and one for the contracting, whereby the son would not be a beneficiary in the contracting business?

That is a common enough situation, and a trap very many tenants fall into. Suppose, on the other hand, that the son has the urge to travel whilst he is as yet unencumbered with family responsibilities. A year or two working on a farm in New Zealand would be wonderful experience, or perhaps a spell working with VSO in a third world country. Sadly, however, time spent overseas, no matter how worthwhile, does not count in the calculation of the 'five years out of seven' rule of eligibility. Quite illogical you may say, and I would agree with you, particularly as a period of three years at University studying a subject, which may have nothing whatsoever to do with farming, does qualify. Illogical it may be, but it is nonetheless the law and yet another trap to be avoided.

There is, however, one way around this particular dilemma. There is nothing in the act to prevent the three years' further education being spent, at, for example, an Australian university. The act does not specify further education in Britain and the working experience overseas can be gained during the long vacations.

Finally, with our hypothetical example, let us assume that the worst happens. Hale and hearty father may have appeared to be, but, unexpectedly, he has a heart attack and dies. Very sad and he will be mourned by all in the district—but at least the farm is secure. The son, now aged twenty-three, qualifies on all counts so he can succeed his father. Again, wait a minute! Are we quite sure? What are the terms of father's will? He may well have wanted to ensure that his two daughters benefited in equal share with their brother. And, of course, what about financial provision for his widow? The fact is that this part of succession planning bristles with difficulties and pitfalls. The point, however, is quite clear. The nominated successor must be seen by the Agricultural Land Tribunal to possess adequate finance to run the business. And I do mean possess. It is not sufficient, for example, for the two sisters to say, no matter how sincerely, that they intend to leave their share in the farm. They may change their minds! The money would have to be bound in by the terms of the will so that they could not take it out, even though they might quite properly benefit from it.

So the wording of the will is vitally important and, if succession is to be achieved, some compromise on other and perhaps equally valid objectives in the planning of the family's future may have to be made. The fact is that it is far from unknown for succession to be lost on these grounds alone and often quite unnecessarily.

SOME CHANGES IN THE 1986 AGRICULTURAL HOLDINGS ACT

Whilst the Act was not in any way retrospective with regard to succession, certain changes were made in the rules. Although most of these were to the benefit of the tenant, they once more emphasise the need for continuing vigilance in seeing that planning is updated. The most important of these changes are:

Retirement

Previously, succession could not take place until the tenant died. The sight of sons hanging on, abiding by all the succession rules, waiting for fathers to die, was not particularly edifying! This has now been changed, and a tenant can opt to retire at the age of sixty-five in favour of a nominated successor. But, as you might expect, it is not entirely straightforward. If mutual agreement is reached with the landowner, well and good. If, however, it is contested, then the inter-vivos succession, as it is called, has to be tested before the Agricultural Land Tribunal. As in all succession cases, the candidate has to prove his eligibility and suitability. It is important to note that, should he fail to convince the Tribunal, he will never again be eligible for consideration even at the tenant's death. So there is only one bite at the cherry! If you are going for this option, be sure that your case is watertight. There is, however, one escape route: if during the hearing, you sense that you may lose, you can withdraw right up to the point when the Tribunal decides. In that case, the application is cancelled and the tenancy continues as before.

There is another possibility to consider. The exclusion of one potential successor does not preclude other possible candidates from consideration either at retirement or at death.

It may be wondered what is so significant about the age of sixty-five. The answer is nothing more than that it is the current national pension age for men, and there is no real logic in insisting on this age for retirement succession. It is difficult to see why a tenant should not be allowed to retire in favour of a successor at any age. It would get rid of one succession, bring younger men into farming and be in tune with current thinking on earlier retirement. This is one improvement that we must seek to gain, but that is for the future.

There is one circumstance where a tenant may opt for earlier retirement, and that is in the case of ill health. If it can be shown that he is suffering from a permanent illness that prevents him from farming in accordance with the Rules of Good Husbandry, then an application may be made at any age.

Limitation to One Commercial Unit

It is frequently the case that a tenant occupies more than one holding either under the same landlord or even several different ones. Succession by one applicant can only be to one commercial unit. What then is a commercial unit? It is 'a unit of agricultural land which is capable, when farmed under competent management, of producing a net annual income of an amount not less than the aggregate of the average annual earnings of two full-time agricultural workers aged 20 or over.'

Clearly, this is important as, when multiple holdings exist, they risk being broken up at the death of the tenant. One successor will succeed to one commercial unit, and he will no doubt choose the biggest and best, but the rest will go. If you are in this position, you should give special thought as to how to manage it. Should you have more than one potential successor who is both 'eligible and suitable', then a certain amount of skilful planning can lead to succession on more than one commercial unit, and maybe all of them. Using the retirement provision, the tenant can get one succession out of the way. He is then no longer tenant of that particular unit and can consider succession planning for the next unit either at a subsequent retirement application or at his death. And so on until all are passed on to the next generation. But be sure to get your planning done correctly and, above all, do not die too soon after the age of sixty-five!

Widows

Under the 1976 Act, it was very difficult for a widow to succeed unless she could prove that she had taken an active part in the actual working of the farm. The 1986 A.H. Act has remedied this and, provided of course that the other requirements of eligibility and suitability are met, then the livelihood test is met by considering her as working jointly with her deceased husband. Note, however, that in the event of the woman being the tenant, this provision does not apply to the widower.

Succession Rent
Whereas under the 1976 Act, the succession tenancy was a brand new tenancy in the sense that the rent was totally based upon the open market, i.e. tender basis, this is now changed so that the new rent is based upon the sitting tenant basis as determined by the 1986 A.H. Act.

Notices
As with many other parts of Agricultural Holdings legislation, strict rules are laid down as to the serving of relevant notices, and the time scale over which these operate. Some of these were changed in the 1986 A.H. Act. Beware, and be sure that you meet these rules. No latitude is allowed.

A FINAL THOUGHT

Throughout this book it is apparent that I disapprove of the succession provisions granted under the 1976 Act. The balance of justice between landlord and tenant was disrupted and damage was done to the system as a whole. Furthermore, in almost every case, retrospective legislation is both unfair and thoroughly undesirable. Anyone in a long-term business such as farming must be given the confidence of knowing that his decisions made today under current law will not be torpedoed by subsequent retrospective legislation.

The Labour Party has on occasion threatened to 'go back to 1976' if returned to power. I think there are good reasons for believing that wiser counsels would prevail, and I feel certain that the TFA would do its utmost to prevent such a harmful move backwards.

Having said that, succession is on the statute book for some if not all tenants. That being so, those tenants should know how to make use of it.

CHAPTER 8

Tenant's Improvements

When dealing with the historical development of the landlord and tenant system in Chapter 2, I traced the beginnings of compensation to tenants back to the great landowner improvers. They realised that their investment was not going to be matched by similar effort by their tenants unless those tenants were confident that they would eventually get the full benefit of their investment. The Lincolnshire Custom followed this and laid down an accepted basis for reimbursing tenants at the end of the tenancy for the unexpired value of their improvements. Such common sense is now enshrined in law, but, as in most matters of this kind, neither reward nor compensation come automatically. And, as ever, the tenant must be wide awake.

The sensible starting point is at the beginning of the tenancy, when a schedule of condition should be agreed. This is quite simply a record of what there is and what condition it is in. This applies to buildings, of course, but also to the land and whether it is dirty, the condition of hedges and ditches and the state of the drainage. It is important to have this record, not only to act as a base level against which improvements may be measured— whether carried out by the landlord or the tenant—but also as the line below which dilapidations may be valued.

If for any reason such a schedule of condition was not drawn up at the beginning of the tenancy, then you should approach your landlord to get one done right away. It is difficult to imagine why he should not want to do this jointly with you; it is,

51

after all, in the interest of both parties. However, there is nothing whatsoever to stop you getting this done by your valuer as a professional record, should the landlord's consent not be forthcoming.

CATEGORIES OF IMPROVEMENT

So we have a point of reference from which to start. Built on top of that base line will be the improvements carried out by the tenant. For the purposes of compensation, a distinction is made between those carried out before 1 March 1948 and subsequently. As 1948 recedes into the past, there will be fewer and fewer instances of such compensation. Nevertheless the distinction must be noted for, to some extent, the method of compensation is somewhat different. For all post-March 1948 improvements, compensation falls into three categories:

1. Those where the consent of the landlord is required, but where there is no appeal to the Agricultural Land Tribunal if consent is refused. Examples under this heading include the planting of orchards, warping, making watercress beds and some other somewhat unlikely enterprises such as the planting of osier beds. Thus, in this category, if you do not get the landlord's consent, it is quite simple: you do not get any compensation.

2. Those where the consent of the landlord is required or, failing that, the approval of the Agricultural Land Tribunal. This is a much larger group of improvements, and includes many which would be considered as the normal type of tenant's or landlord's improvements: drainage, permanent fencing, buildings, roads and so on. Clearly it is much more satisfactory if you can get the landlord's consent (in writing, please), but these improvements are recognised as such important tools of the trade that the ALT is there as a

backstop. You should note that appeal to the Tribunal is possible even where the landlord has given consent but only on what are considered unsatisfactory terms. In giving their decision, the members of the Tribunal will therefore lay down the terms that they consider fair and just; they will also normally give the landlord the option of doing the work himself.

3. Those where no consent is required. These are basically of a short-term nature and are mainly concerned with the husbandry of the farm. Liming, fertilising, consumption on the holding of feeding stuffs, both purchased and home grown, are some obvious examples. Consent may not be required, but accurate record-keeping most certainly is. Receipted invoices of all the materials purchased and work done are a basic minimum.

These then are the three categories under which tenant's improvements are dealt with. Of course, improvements do not necessarily have to be done by the tenant. It may well be that the landlord will choose to do them himself, or at least offer to do so. If given the choice, the tenant must decide which is the better option, depending upon cost, and whether it is cheaper to use his own money or to pay the landlord a percentage, which is often less than bank lending rates, which will be added to the rent. In fact, many landowners and their agents prefer to fund improvements rather than build up what might become a substantial liability at the end of the tenancy.

RATES OF COMPENSATION

Assuming, however, that the improvement in question is to be paid for by the tenant, on what basis should he eventually be compensated? The tenancy agreement may have something to say about this, but in many cases it is a matter for negotiation.

The exceptions are mainly to be found amongst the short-term improvements where agreed valuation rates are laid down as unexhausted manurial values.

It is with the medium- and longer-term improvements that we run into some problems. With a building, for example, both theoretically and practically it is perfectly possible to value its worth to the farm at any given date. Indeed, the 1986 A.H. Act states the principle that the value shall 'be an amount equal to the increase attributable to the improvement in the value of the agricultural holding'. That seems perfectly fair; for example, a building which cost £10,000 to build in 1975 and is still in excellent condition in 1988 may well be worth perhaps £20,000 at current values, and indeed would cost far more than that to replace.

However, a custom has grown up whereby many landowners seek to impose on their tenants terms which write the value of the improvement down to a nominal figure, usually £1, after a determined number of years. There really is no reason why a tenant should accept an unreasonably short term for a long-term improvement, or indeed any term at all. Nor any reason why the value should come down to £1. The fact is that many tenants have been bluffed into accepting such unfavourable, indeed unjust, terms. As always in these cases, the lesson is perfectly clear. Look before you leap! But it is also worth considering whether any purpose is served by a retrospective appeal to the ALT. It is impossible to say without particular knowledge of the circumstances, but it is well worth getting advice.

These then are those matters which must be dealt with during the tenancy. Other matters concerning compensation, which only arise at the end of the tenancy, will be dealt with in Chapter 11.

CHAPTER 9

Responsibilities and Obligations

Right from the earliest days of the landlord and tenant system, one theme has stood out above all the others: that is, that the land should be farmed in a 'goode and husbandlyke manner'. The realisation that the extractive farming of the manorial system of strip farming impoverished the land, along with the need to convert that farming from subsistence to commerce in order to feed a rapidly growing population—these two were the factors which fired the imagination of the great improvers and breeders, and the developing system of land tenure was, if you like, the fireplace on which that fire burnt and flourished. Inherent in it was the responsibility to farm not just for today but for tomorrow and the day after. That responsibility was finally put on the statute book in the 1947 Agriculture Act in the form of the Rules of Good Husbandry and the Rules of Good Estate Management.

All the other duties laid upon the tenant are to be found in the tenancy agreement, and it is to that which we must look in each individual case. Bearing that in mind, let us look at some examples.

1. Cropping and Stocking
Many agreements contain references to the need to maintain a rotation, not to plough up permanent pasture and even to 'spud thistles' every year. These are largely a carry-over from the days before the arrival of modern agricultural technology. But we

should not pass them over and ignore them. Some may well be of importance, and carry penalties if ignored. Take permanent pasture for instance. Many farms include an area of parkland, the more important if it lies in front of the landlord's house. But there may be good farming reasons for wanting to plough it up. It is perhaps only suitable for sheep and has become sheep sick and needs a rest. Amenity pasture at a low rent in front of the mansion is one thing; old pasture of no particular amenity value and at a high rent is very much another. The Act provides for relief in these cases, enabling the tenant to demand a reference to arbitration seeking permission to plough. Other clauses referring to cleanliness of the land, e.g. freedom from couch, wild oats, black grass, etc., are very important in that if the land is dirty, very substantial dilapidations can result at the end of the tenancy.

2. Repairs and Insurance

Repairs and insurance are the commonest of all the obligations laid upon the tenant. The 1986 A.H. Act provides for certain basic minimum requirements to be mandatory in the absence of any clause in the agreement to the contrary. These are usually known as the Model Clauses and they are to be found in Statutory Instrument 1973 No. 1473, not in the act itself. Briefly, they provide for:

- the landlord being responsible for main structure and roofs, external decoration and all underground water supplies
- the tenant being responsible for the rest, that is to say, all interiors, plus fixtures and fittings; and, on the land, for fences, hedges, gates, watercourses, drains, etc.

This is a commonsense division of responsibility which has stood the test of time. In retrospect, one can say that it had one particular advantage. It brought the agent, either resident or from a local firm, into regular contact with, and responsibility for, important parts of the farm's fixed equipment. It gave him something of fundamental importance to do other than being a mere rent collector. And it brought him into close contact with the tenant.

The modern development has been to shed this responsibility and to pass the whole obligation for repairs and insurance over to the tenant. This is the so-called 'fully repairing and insuring' agreement—FRI for short. This change has largely coincided with the move away from the resident agent towards the use of large and mainly City-based firms of agents. Too busy, often too distant, and seeking economy, they have shed this load. Whether this is the case or not, it is certainly true that the FRI agreement simplifies estate management in that it passes both responsibility and cost on to the tenant. This may, or may not, be a good thing but it is my personal opinion that such agreements are a major mistake in that they profoundly alter the role of the landlord's agent.

Be that as it may, full responsibility for repairs and insurance, involves a cost, and a substantial one at that, to the tenant, and he should certainly not take it on—if he has any choice in the matter—without first doing his arithmetic with great care. Logically, the rent under an FRI agreement should be significantly less than under a Model Clause agreement. Just how much depends on many things. How old are the house and cottages? In what state of repair and condition? How many buildings are there? Are they modern, low-maintenance buildings, or are they old and fragile? It can make a very big difference.

On many farms, there are buildings which are really not worth spending money on, either for repair or for insurance. In those cases, it is well worthwhile discussing with the landlord the possibility of having them declared redundant; neither party is then responsible for them and no obligation exists. Do, however, make sure that the buildings are not listed before taking this step.

3. Residence in the Farm House
Many agreements contain a clause requiring the tenant to reside in the farm house. It is easy to understand why—the tenant is on the spot and will therefore take a keener interest in the general upkeep of the property. An absentee tenant, being at a distance, may well take a much more detached view.

I can see nothing in principle to object to in this, but, in practice, it can cause difficulties in the period running up to retirement. For one thing, if the tenant is planning to build himself a house to which he can go at sixty-five, he will not be able to get tax relief on a mortgage on what is regarded as a second house. And by the nature of things, his qualified successor (if he has one) will be likely to have a young family whilst he and his wife would no doubt welcome the chance of moving to a smaller house rather earlier.

Residence in the farm house is thus an obligation which could be dealt with rather more sympathetically than is sometimes the case.

4. Stone Walls and Banks
Whether they be Cotswold stone walls, Cornish earth banks or stone dykes in the hills, these can be a real problem. Built in an earlier age when field size was much smaller, many of them are redundant and have fallen into disrepair. Others have, of course, been well maintained and serve a useful purpose. The point, once more, is what does the tenancy agreement say, and what record is there of condition at the start of the tenancy? If there is an obligation to maintain in good order but no record of agreed redundancy, then you could be in for a very nasty surprise. Some dilapidation claims in Cornwall involving complete restoration of banks have left the outgoing tenant in a very unpleasant position indeed.

NOTICES TO REMEDY

It is one thing having a list of obligations in the agreement, but what powers exist to compel a tenant to fulfil his responsibilities if he chooses to ignore them? The answer is that a landlord has the right to serve on the tenant what is known as a Notice to Remedy, which can be used when the landlord considers the tenant to be in breach of the terms of his tenancy.

There are two forms, each for a different purpose, and it is essential that the correct form is used, as otherwise the notice is invalid.

Form 1 is that requiring the tenant to carry out works of repair and maintenance as specified in the Tenancy Agreement or under the Model Clauses. He will be given three months within which he must carry out the work. Within one month of receipt of the notice, the tenant may demand arbitration in respect of either his liability, the reasonableness of the work required or the time allowed. If that step is not taken, and the landlord then serves a Notice to Quit at the expiration of the time allowed (as he has the right to do), then, once more, the tenant can demand arbitration within one month—but only on the basis of challenging the reasons stated in the Notice to Quit. Finally, should the tenant fail at arbitration, he can refer the Notice to Quit to the ALT. The tribunal will only give its consent if it is satisfied that a fair and reasonable landlord would insist on possession.

Form 2 is that requiring the tenant to comply with terms of the tenancy agreement other than for repairs and maintenance. It covers such matters as occupation of the farm house, prohibited sales of grass keep, etc. The tenant must carry out the work within a 'reasonable time', and not more than six months. Arbitration, which is not available to the tenant following a

Form 2 Notice in the first instance, is only available if failure to comply with the notice is followed by a Notice to Quit. Subsequent reference to the ALT is not allowed.

It should be noted that it is not only the tenant who has obligations and responsibilities; so too does the landlord. The tenant has an equal right to serve Notices to Remedy where these are justified.

Let us recognise that the serving of notices must be regarded as a last resort, an admission of some sort of failure. Every effort should be made by both sides to resolve difficulties long before formal notices are served. It is a sad reflection on the state of the relationship should they become necessary. But if that stage is reached and a tenant receives a Notice to Remedy, then he should be off to consult his solicitor *at once*. The time limits are short and absolutely binding. The tenant must beware that failure to comply, or to appeal successfully, can lead to the loss of the tenancy.

CHAPTER 10

How to Get a Tenancy

Right through the 1970s and early 1980s, tenancies were only obtained with very great difficulty. As I have said elsewhere in this book, the supply of farms to let largely dried up, particularly following the 1976 Act, as landowners sought to keep access to vacant possession. At the same time, following Britain's entry into the Common Market, agriculture passed through a decade of high prosperity. All the conditions, therefore, leading to an excess of demand over supply. It was not surprising that it was almost impossible for an aspiring tenant to find a farm to rent, nor that those few farms that did come on the market were let at very high tender rents. The position has, however, eased somewhat since the 1984 Act became law and two-generation succession was abolished; this also happened to coincide with a marked downturn in profits as a result of efforts to control surplus production.

So how does someone set about getting a tenancy? Either by personal contact or by application following an advertisement.

The former, it goes without saying, is by far the more certain route but, by the nature of things, tends only to be available to farmers' sons living and working on the same estate or, at the very least, to someone from that particular district. Who you know is of absolutely critical importance, and it is very easy to understand why. We are not talking about a 'here today, gone tomorrow' relationship, and confidence in the tenant as a person and in his family will influence the choice of a landowner or his

61

agent. Especially is this true on the great landed estates and where there is a resident agent. Here concern for the country-side and the recreation to be found locally, perhaps hunting or shooting; an interest in the Church and local school; and a reputation for solid worth and reliability—all these charac-teristics will be taken into account, along with the applicant's farming ability and his financial resources.

This sort of farm on this kind of estate will almost never appear in an advertisement. But there are other classes of landowner, be they institutions like the old-established ones, such as the Crown Estates, the Church and the Oxford and Cambridge colleges, or the newcomers, pension funds and the like. And then, at the lower end of the acreage scale, there are the County Councils with their smallholding estates. The advantage of personal knowledge and contact is less important with this kind of landlord, though it would be wrong to say that it does not count. The approach is much less personal and much more a matter of straightforward business judgement.

THE APPLICATION

The advertisement will generally give an outline of the qualifications required: what sort of age range, how much previous experience and how much personal capital is needed. Detailed particulars of the farm will usually be available on the payment of a fee, and these will sometimes include an appli-cation form. If so your plan of attack is made that much more simple in that the required format is clearly set out.

Whatever the case, your application must be fully set out, but at the same time be concise and follow a logical pattern. It should include:

1. Personal details of yourself and your family, including your background and your particular interests.

2. Details of your technical education and farming experience.
3. Your plans for farming the holding with the proposed cropping and stocking, together with your ideas of how to deal with any particular problems or opportunities.
4. Details of the capital you have available, plus your access to further finance from the banks. It will be essential to show that you have sufficient funds in order to be able to farm properly, and also to survive any unforeseen difficulty.
5. You will be expected to tender a rent, and this will need to be supported by a financial budget showing how you justify the figure.
6. A list of suitable persons to whom the agent may make reference. These should include a banker, as well as someone who can testify to your farming ability.

You will, of course, be able to visit the farm and you should take the opportunity to walk every field. You should certainly take a spade or a soil auger with you so that you can have a good look at the soil type, the sub-soil and the general condition and structure of the land. If the farm is drained, then be sure that you get a copy of the drainage map so that you can look closely at its effectiveness and the condition of the outfalls and so on. You will need to inspect the buildings very closely, especially so if the tenancy is to be based upon the tenant taking full responsibility for repairs. Then, in addition to the inspection, do your best to find out what you can from the locals; a bit of pub gossip isn't out of place, and a chat with the local ADAS adviser will certainly be worthwhile.

THE TENDER RENT

You have got all your ideas assembled, your budget is prepared and you are ready to work out just how much you can afford to pay. You will also have a copy of the tenancy agreement which you will be expected to sign. The first step, before you do anything else, is to get both your valuer and your solicitor to go through the agreement with you, clause by clause. It is important that you should seek the advice of both professionals so that you are fully aware of all the implications, all the obligations and the legal interpretation of them. Find out what ingoings you are expected to pay for: what level of tenant right, which fixtures and, if there is a milk quota, whether you have to pay for it and how much. You will need also to consult your accountant and, with him, to get the support of your bank.

When you have taken all this advice, you come back to the crunch question—how much can you afford to pay? Unless economic conditions worsen a great deal, and the competition for farms fades away, it is certain that if you are to stand any chance of getting the farm, you will have to pay more than you would wish. There will also be the element of premium or key money.

The best way to approach this, I believe, is to separate the economic rent from the key money. Do your budgets realistically, and most certainly not over-optimistically, in order to arrive at a rent which makes sense. Use the productive and earning capacity formula that you would use in a rent review and arrive at a calculation of the cake. Work out how big a slice you can afford to give away. Then, and only then, add on the premium which you are prepared to cough up in order to get the farm.

I believe that most agents respect this approach. Very few will be so stupid as to accept the highest tender, because they will be looking for a responsible tenant whose future is viable and who will be able to farm properly. They will be impressed by the practical and realistic approach to your budgeting, and they will know that you will use this approach in future rent reviews.

There is a further argument in favour of this split between rent and premium. If you tender a single figure combining the two, then you will be stuck with it in future. Rent review negotiations and possible arbitration will be much more difficult if you have to try to go down from that single figure. Separate the two, and you have a better chance of coming back to reality in three years' time.

A final word. You want to farm, and you are absolutely confident of your abilities. You are in perfect health, and so is your wife, and you are both prepared to work all the hours there are to realise your ambition. Don't let me for one minute try to quench your enthusiasm, but do remember—things can go wrong; illness can strike; the weather can be impossible; market prices can collapse. Don't go so overboard that you cannot cope with the unexpected. If you lose the farm to someone else who, for some reason, is able to bid a ridiculous figure, then so be it. Far better to lose it and try again elsewhere than to saddle yourself and your family with a dangerously heavy burden.

SHORT-TERM TENANCIES

These and other forms of alternatives to a standard tenancy are dealt with in some detail in Chapter 14. However, the point is worth making here that some of these arrangements may well be of particular importance to the new entrant who lacks the necessary experience which would commend him to a landlord. This is especially true with the Ministry five-year licence scheme. Be that as it may, the application procedures logically follow the same pattern, although my final warning of the dangers of over-commitment will need to be read in a rather different context.

CHAPTER 11

The End of the Tenancy

All things, good or bad, come to an end eventually; and a contract of tenancy is no exception. We have to be sure that, when it does, the process is managed correctly. There are, of course, two ways by which a tenancy can be terminated: either by the action of the landlord seeking successfully to get vacant possession, or by the tenant granting it, voluntarily or involuntarily (in the case of death).

NOTICE TO QUIT

As we have seen in Chapter 9, when considering the tenant's responsibilities and obligations, a Notice to Quit can follow a Notice to Remedy in the event of the tenant failing to comply. There are, in fact, a whole list of circumstances which can lead to a Notice to Quit being served. They divide into two different classes: incontestable and contestable. At the risk of stating the obvious, an incontestable Notice to Quit means that the tenant has no recourse to an appeal against it, whereas a contestable notice is one where the tenant has the right to serve a counter notice.

GROUNDS FOR AN INCONTESTABLE NOTICE TO QUIT

It is important for a tenant to be well aware of just what circumstances can lead to the serving of an Incontestable Notice—if only for the very good reason that some of them can be avoided. They are known as the Seven Deadly Sins, although there are in fact eight of them, and, briefly, they are as follows:

Case A   Where there is a Statutory Smallholding, in which case the County Council has the right to serve the notice when the tenant reaches the age of sixty-five, and it is effective subject to the condition that satisfactory alternative accommodation is made available

Case B   Where the land is required for a use other than agriculture and where planning consent has been obtained or is not required

Case C   Where the ALT has granted a certificate that the tenant is not fulfilling his obligations under the Rules of Good Husbandry

Case D   Where the tenant has failed to pay the rent due within two months of receiving a notice requiring him to pay; and/or where, within a reasonable period, he has failed to remedy a breach in the terms of the tenancy agreement which is capable of being remedied

Case E   Where, at the date of serving the notice, the interest of the landlord is being materially prejudiced by a breach which is not capable of being remedied, e.g. bad husbandry causing lasting deterioration, or gross neglect of buildings

Case F   Where the tenant is insolvent

Case G Where the tenant has died. The notice must be served within three months of the death being notified to the landlord

Case H Where the notice is served by the Minister of Agriculture in order to enable him to use the land for any amalgamation or reshaping of the agricultural unit

It goes almost without saying that an incontestable Notice to Quit is bad news for the tenant, with the exception of the County Council smallholder, who has of course been expecting it and has been planning for his retirement, and also in the case of development, where there will be some financial benefit arising from what is generally only part of the holding. So the first rule must be—do not lay yourself open to such a notice. And the second rule is—if you receive one, act and act fast. Pick up the phone and be with your advisers today rather than tomorrow. There are strict rules governing the serving of counter-notices without which the right to claim compensation is lost.

A word needs to be said about the circumstances following the death of the tenant. Very often I have heard criticism of the insensitivity of a landlord when a Notice to Quit has been served indecently soon after 'old so and so has been laid to rest'. I can sympathise with the point and it certainly doesn't hurt anyone to do what has to be done in a kind and civilised manner. Having said that, it must also be said that life has to return to normal as soon as reasonably possible. Normal in this instance means what the law lays down, and to be effective, the Notice to Quit must be served within three months of receipt by the landlord of the notification of death from the deceased's executors. That has to be done whether occupation by the family is to come to an end or whether succession is a possibility. In either case, the landlord's advisers dare not risk invalidating the notice as a result of it being late.

CONTESTABLE NOTICE TO QUIT

These are quite simply those not covered by the Seven Deadly Sins, and we have already dealt with the most important of them under the tenant's obligations to maintain and repair. In order to contest, the tenant will need to serve the appropriate counter-notice, requiring the matter to go to arbitration and/or the Agricultural Land Tribunal. As I keep saying, don't sit around if one of these notices should drop through your letter-box.

COMPENSATION ON THE TERMINATION OF TENANCY

This can work both ways: the tenant expecting to be paid for disturbance, for improvements and for tenant right, whilst the landlord will be claiming payment for dilapidations.

1. Compensation for Disturbance
This only applies under certain circumstances, and certainly not at what one might call the normal end of a tenancy. A claim can only be made when termination is due either to a Notice to Quit from the landlord or a Counter-notice to Quit the whole of the holding following a Notice to Quit a part. Compensation now falls into two parts: basic and additional compensation.

The basic compensation will amount to one year's rent, or a greater sum calculated on the basis of the actual loss but not exceeding two year's rent, provided the tenant has made the appropriate claim not less than one month before the termination date.

The additional compensation is that sum needed to assist the tenant in the reorganisation of his affairs, and may be equal to a maximum of four years' rent.

Thus the total compensation which a tenant would get is

limited to the equivalent of six years' rent. In the case where the land is being developed for housing and the value becomes many thousands of pounds per acre, a maximum of six times, say, £50 per acre, may seem to be somewhat niggardly—but that is how it is.

2. Compensation for Improvements
This area has been dealt with in Chapter 8 under the general title of the tenant's improvements.

3. Compensation for Tenant Right
This covers payment for growing crops, cultivations and the unexhausted manurial values of fertiliser, manure and lime. It also includes somewhat less obvious 'benefits' left on the farm such as the acclimatisation, or hefting, of hill sheep on to a particular hill farm. All this is a matter for a skilled valuer following the accepted codes of practice laid down by his profession. You should note that, legally, the negotiations take place as between the outgoing tenant and the landlord. Very often, in practice, where there is an incoming tenant, his valuer will take the place of the landowner as it is he, the incomer, who will be paying for the tenant right.

4. Compensation for Dilapidations
The landlord is entitled to claim compensation for any dilapidation, deterioration of, or damage to, the holding as a result of the tenant not fulfilling his responsibilities to farm in accordance with the Rules of Good Husbandry. Furthermore, any contractual obligations entered into by the tenant in the tenancy agreement but which have not been met will be included in the claim. Thus, the state of repair of the house, buildings and fixed equipment, depending on whether the agreement is based on Model Clause or FRI, will come into the calculation. In addition the claim will cover such things as neglect of fences and gates, roads, ditches and drainage, dirty land (couch, black grass, wild

oats, etc.), irregularity of cropping and even the selling off of produce contrary to either custom or agreement.

It is quite clear, therefore, that the valuation of the dilapidations claim can be very wide-ranging. The basis of the claim will be that of making good the dilapidation or damage. For example, in the case of neglected drainage, stone walls which have fallen or hedges which have been pulled out without consent, the amount of money involved can be very considerable indeed. This may seem to be a completely open-ended arrangement, and indeed it was prior to the 1984 Act. But in that act, a clause was included which was designed to limit the amount of compensation payable: 'it shall in no case exceed the amount, if any, by which the value of the landlord's reversion in the holding is diminished owing to the dilapidation, deterioration, or damage in question'.

This limitation set on dilapidation seems like good news, and indeed it is, for a Cotswold farm with many fewer stone walls than thirty years ago is worth no less, and may even be worth more, as a result. But it must be said that the good news is tempered by the difficulty of valuation. How exactly does one assess the amount by which the value of the landlord's reversion has been diminished? A fertile ground indeed for lengthy argument between the valuers!

CHAPTER 12

Guide to a Tenant's Essential Records

If you have stayed with me this far, you will surely be well aware that the law is complex, that there are pitfalls galore into which you need not fall and that, in very many cases, you need to take action quickly and concisely. It should not, therefore, be necessary to say that your record-keeping should be both complete and well organised. It is surprising, not to say alarming, how many tenants do not know where their tenancy agreement is (on the mantelpiece behind the clock may be more than a music hall joke!). If for no other reason than that it could cost you a lot of money, do get your records both complete and in order.

Before I attempt to set out a list of what I think is a minimum, let me repeat once more the plea that you should employ a valuer on a regular basis, and that he should work closely with your accountant and solicitor. Provided he is good at his job, then you will be at least halfway out of trouble. During the process of doing the annual valuation, which will form part of your accounts, he should be assembling much of the information which would be needed, for instance, in working out claims for compensation. But having such a person at your elbow does not in any way absolve you from providing him with the records and the evidence that he requires.

1. The Tenancy Agreement
This is the vital document, and do not let it gather dust. It needs to be kept up to date by means of a legally endorsed

memorandum each time any condition of the agreement is changed. Changes in the rent, for instance, every three years, are obvious, but less so is the necessity to record changes in the schedule of tenant's fixtures and redundant buildings; or in the schedule of acreages, particularly if these have been changed by, say, the creation of a conservation area, or if some parts of the farm are taken out by agreement for tree planting. Not attached to, but certainly kept close by, the tenancy agreement should be the detailed plans of field drainage and water supply. Landlord and tenant relations apart, how often is it that when a wet spot develops in a field, you can go straight to it and find both the outfalls and the line of the drains?

2. Rent Reviews

Remember that you are going to have to deal with the productive capacity of the farm and its related earning capacity. You may well have to contest figures put forward by the landlord's valuer, and you may be sure that he will be setting his sights on the high side. What evidence have you got of actual yields and stocking rates? An absolute essential is a year by year record of cropping and stocking, along with yields both physical and financial. Preferably these should cover the whole period of occupancy of the farm, but they should go back for five years at the least.

Furthermore, you will need to be able to budget forward to justify your claim that the future rent should be such and such a figure. Remember that the arbitrator's award will be based, partly at least, on what the hypothetical competent tenant can achieve—so your figures must be assembled in such a way that they are comparable with university, Ministry, MMB or MLC costings.

Then you will need to have records of both landlord's and tenant's improvements—a full description, the date when the work was carried out, and how much it cost, including details of grants and subsidies.

I made the point in the chapter dealing with rent reviews that the preparation of your case needs to be based on fact, and carefully documented fact at that. Pub gossip just will not do. Your case is three parts of the way to being lost if you do not have the documents and records to hand; and the cost of preparing your case by your valuer will be enormously greater if he has to dig and hunt for what he needs.

3. Tenant Right
Again, complete cropping and stocking records will be needed; but, in addition, invoices of fertilisers, lime, feeding stuffs and seeds covering, in some cases such as lime, the previous eight years. These details *ought* to be included in your annual valuation, and if they are, so much the better.

4. Disturbance or Damage
It may be the result of a new pipeline, electricity power lines or the building of a new road. There will be loss and damage whilst the work is in progress, and it must be meticulously recorded step by step. This is the point to say that every tenant should have a good camera. Really good and clear photographs are an almost essential part of a successful claim. Incidentally, one piece of advice which is relevant to pipelines where there is an excavation of the subsoil is to never sign a release document, because you can never be sure how the land will react afterwards. One pipeline put in on my farm over thirty years ago still shows up in significantly lower yields.

5. Tenancy Succession
As we have seen in Chapter 7, succession is not automatic, contrary to what many people believe. No tenant—no matter how young, no matter how good your health, no matter how long before you envisage passing on the farm—should neglect planning for succession. When the time comes, your solicitor and your valuer will need a great deal of information, and it may

be, to put it bluntly, that you will be dead, and in no position to help! Your advisers will have to move swiftly and it will be so much easier for them if they are brought in long before the event. Expensive? Yes, but not disproportionately so, and, in all conscience, can you afford not to?

A FINAL WORD ON RECORDS

Reading through this chapter, it seems to me to resemble the nagging of an old nanny! It is all a bit of a bore, and, like you, I would much rather be outside than stuck in the office. Of course, if your records are all in apple pie order, and you know exactly where everything is, it does not apply to you, but I am prepared to bet that you are in a very small minority of farmers, tenant or otherwise.

CHAPTER 13

Quota Complications

To describe milk quotas as a complication has to go down as the understatement of the century. Just when we thought we had begun to get the landlord and tenant system moving back to some semblance of normality by undoing part of the damage caused by the 1976 Act, along came milk quotas in April 1984 to throw a very nasty spanner in the works.

We seem to have forgotten that milk is not the only product subject to quota control. Both sugar beet and potatoes have been controlled for a long time, and it is perhaps all the more surprising that such a mess should have been made with milk quotas when there were successful examples to follow close at hand. It was, I think it is fair to say, the result of decisions taken in Brussels by people who had no clear understanding of how our unique system of land tenure worked.

Before coming back to the specific problems of milk quotas, I must deal with one aspect of quotas generally. Many tenancy agreements, but by no means all, contain a restrictive clause whereby the tenant is required to 'use his best efforts' to maintain a quota on the holding. There is no one consistent wording which is used—the TFA has identified many different variations on the same theme—but the sense is the same. The tenant must do his utmost to maintain, or is forbidden to dispose of, any existing quota on the farm. The implications of this type of clause are quite worrying as disposal of quota could clearly lead to a claim for dilapidations at the end of the

77

tenancy. But what if conditions have changed, as they have most certainly with potato growing. Mechanisation, irrigation, and climatically controlled storage—all these have changed the face of potato technology, and without them, and certainly without suitable (i.e. easy to mechanise) soil, the one-time potato farm might well be better off growing some other crop. So there have been potential difficulties, but I think it would be accepted that the system was sufficiently flexible and possessed of commonsense to carry the problem until milk quotas came along.

The difficulty started with the attachment of the milk quota to the land rather than the producer, and, at the same time, the Ministry of Agriculture compounded the problem by so organising the quota attachment that it assumed a cash value. Overnight the tenant's capital was depreciated as the value of cows fell, and of course his income potential dropped as well. The landlord's capital, on the other hand, increased because the quota was attached to his property. We have seen how valuable this is when farms advertised for sale, with a substantial milk quota attached, fetch a higher price than they would otherwise do. It is no exaggeration to say that the relationship between landlord and tenant was soured by this episode. Not surprisingly, the CLA felt obliged to hang on to what had unexpectedly dropped like manna from heaven into their members' laps, whilst tenants felt a deep sense of injustice.

MILK QUOTAS AND THE AGRICULTURE ACT 1986

The 1986 Act set out to try to repair some of the damage and, in particular, laid down a formula whereby for existing tenancies the tenant would be able to claim compensation for his share of the quota at the end of the tenancy. It is a complicated formula and this book is not the place to set it out in detail. Quite apart from its complexity, it will no doubt change, and so anything that I write today could quite easily be out of date in a very short time.

There are, however, some points of fundamental importance which should be stressed. The overall quota on the farm is split into two parts, the landlord's share and that 'belonging' to the tenant. I say 'belonging' because strictly speaking it does not belong to the tenant until the end of the tenancy, when he may receive compensation for that part of the quota which can be agreed to be the result of his efforts or investment. The consequence of this is that no capital value can be put upon it during the tenancy and therefore it cannot be used by the tenant as collateral to secure his bank borrowing. This was a particular problem when cow values dropped and the tenant suddenly found that his borrowing was insufficiently well covered. As a result of an initiative launched by the TFA in 1987, it is to be hoped that we may be close to an agreed solution to this very real worry.

THE BASIS FOR COMPENSATION

Perhaps of greater consequence is the need to establish the base on which the eventual compensation will be calculated. 1983 will normally be used as the base year, and consequently, the valuer will need accurate and specific records on which to base his apportionment: how fields have been used during the year, whether for grazing or conservation, whether for milking cows or dry stock or whether some parts have been used to grow crops either for feeding or for sale. It is desirable to do this calculation whilst the facts are still clear in people's minds and can be substantiated—note what I said in the previous chapter about the need for records. This will establish the fractions of the Standard Quota due to both landlord and tenant. Note that it does not put a cash value on it, but merely arrives at a percentage allocation of the whole. The actual value can easily be taken from the market value of the quota at the date of termination of the tenancy. Note too that this applies only to the Standard Quota and that the full value of the Excess Quota is due as compensation to the tenant.

No doubt it would be helpful if I defined the difference between Standard and Excess Milk Quotas. The Standard Quota represents what the Ministry considers reasonable in terms of litres of cow's milk that can be expected to be produced from the land in twelve months. The Excess Quota is production above this basic figure.

There are, of course, procedures to be followed. It would be sensible, in my view, for landlord and tenant to get together to agree the fractions of the Standard Quota. This will surely avoid arguments at a later date. However, if the landlord is not prepared to do this, there is nothing to prevent the tenant, with his valuer, doing the work and preparing the case to be used when the time comes. At least they will be speaking with the advantage of recorded knowledge rather than relying on hazy memory. There is, however, provision in the act for either party to serve a notice demanding arbitration in order to determine the fraction. I find it difficult to see the point of doing this. When the end of the tenancy arrives, the outgoing tenant has to serve an appropriate notice claiming compensation within two months of the end of the tenancy. Note that time limit again. If agreement has not been reached within eight months, then the matter goes to arbitration, and the compensation will eventually be paid.

All this, so far, has been concerned with existing tenancies, anticipating them coming to an end in due course. But what of new tenancies contracted since 1986? The outgoing tenant will have had a claim for compensation, and the landlord will have been legally obliged to pay him. However, it is possible, as with tenant right, for the landlord to charge this on to the incoming tenant. What is more serious is that it is also possible for there to be a clause in the new agreement whereby the new tenant agrees to forego his claim to compensation at the end of his tenancy. In other words, he is paying for something which he immediately passes over, and to which he has no title either now or at the end of the tenancy. That seems like a bad bargain to me, and prospective tenants for dairy farms should be on the lookout for this one. It is also right to say that such a clause contracting out of compensation is of dubious legality.

CONCLUSION

The lesson of milk quotas surely is that there really must be a better way to organise our affairs without leading to the damage that has arisen from the overhasty application of quotas to milk. No doubt it will be argued that it is impossible to announce the introduction of quota control—otherwise every producer races to increase production in the meantime. I would like to think that the organisations looking after the interests of both landlords and tenants, the CLA and the TFA, could get their heads together and work out some principles and ground rules that would help to avoid some of the more obvious and painful pitfalls.

CHAPTER 14

Alternatives to a Standard Tenancy

By a standard tenancy, I mean either an annual tenancy or a lease for a number of years, either of which gives the tenant security of tenure for his lifetime (except for County Council smallholdings), and perhaps for another two generations (pre-July 1984 tenancies). Inventive minds have for a long time sought to find ways and means of circumventing security of tenure, with the object of retaining the vacant possession value for the landowner. The danger with such arrangements is that, unless they are set up with great care, they will come to be considered as tenancies at law. Nevertheless, these alternative types of occupation, many of them taking the form of some kind of joint venture, are becoming increasingly popular. There is no doubt that we shall see an acceleration in their use.

1. Grazing Agreements

It is common practice to let what is known as 'grass keeping' specifically for either grazing or mowing for the season, or for some specified period of the year which is less than 365 days. Note that it must *only* be for grazing or mowing. If any cultivation comes into the arrangement, even ploughing and reseeding, there is a danger of conversion to a tenancy. It is a common fallacy that the same immunity from granting a tenancy exists where the land is taken for the growth of an arable crop, carrots for example. That is not so. A grazing agreement is precisely what it says; it is as the 1948 Act states, 'only for grazing or mowing during some specified period of the year'.

2. Ministry Licences

Under the 1986 A.H. Act, the Minister is empowered to grant a licence approving a letting for not less than two years and not more than five, provided both the landlord and the prospective tenant make application to him in writing. Approval is likely to be given where:

- the landlord wishes to give a tenant with limited experience a period of trial before granting a full tenancy
- there is a definite intention that the landlord's son or daughter will take over the holding within five years

In both cases, it must be emphasised that approval is not given automatically and it is certainly no open door for five-year term tenancies.

In addition, a short-term licence can be granted to a specialist grower, which covers the requirements of the carrot, potato or brassica grower seeking clean land.

3. Gladstone v. Bower Tenancy

The name of this kind of tenancy comes from a case decided in the Court of Appeal which ruled that the security of tenure provisions do not apply to agreements covering more than one year but less than two years. Obviously, this type of occupancy can only have a very limited degree of usefulness. Where it is applicable, the opportunity is there, but clearly great care must be taken in drawing up the agreement in order to avoid the risk of granting lifetime security.

4. Partnerships and Share Farming

The essential point in these arrangements is that the occupier (we must not call him the tenant) does not have exclusive possession of the land—that is the distinguishing feature which prevents the contract from being converted into a tenancy with security. It thus follows that it is necessary to ensure it is not a sham partnership, one in name only but without the necessary participation in risk and profit or loss.

What are the differences between a partnership and share farming? Essentially, it is that in a partnership, each partner has unlimited liability for all partnership debts, whereas in share farming, each farmer has his own separate and distinct business. What is common to both, in addition to the lack of security of tenure, is that the relationship between the two parties is not regulated by statute. There is no Agricultural Holdings legislation under which the rules of the game are spelt out. It is a contractual business arrangement between two consenting adults who sink or swim by their own joint efforts. Let me emphasise once more that very great care needs to be taken in drawing up partnership or share farming agreements. I can do no better than refer my readers to the excellent works by Richard Stratton published by the CLA under the titles *Joint Ventures in Farming* and *Share Farming*.

A final point which applies to both share farming and partnerships. You can take all the best advice available and have a perfect contract drawn up, but the success of the business will depend more than anything else on the personal relationship between the two partners. Mutual confidence in the ability and trustworthiness of each other is the essence of success. Without it the best contract will not save the business from disaster. It follows from this that these arrangements are very unlikely to be the means whereby young new entrants will find a way into farming, as no landowner is likely to take this sort of risk with someone whose performance is unknown.

5. Contract Farming

The use of contractors in farming is well established covering a wide range of specific jobs. The justification is normally the ability to use large and powerful machinery on farms which are not big enough to finance the investment on their own. Such contracting can, however, be taken much further and extend to cover the whole farming operation. Obviously this is much easier, and therefore much more common, with arable farming. Livestock pose problems which make them unsuitable for what one might call distant management.

The key is that the owner remains as the farmer. It is his business, he is taking all the risks and he is merely paying someone else to do the work. It is quite common for a neighbouring farmer to take on the complete farming operation in this way—and it is a good way for that farmer to expand his business activity—but there is no security and no risk of creating a tenancy.

THE ADVANTAGES OF JOINT VENTURES

For the landowner the objective and the perceived advantage of a joint venture are quite clear: to create a business relationship with a 'user' of the land which is not constrained by legislation and does not therefore deprive him of access to vacant possession. What is not always so clear is exactly how to achieve this without risking the creation of a tenancy. The essence, therefore, of a successful joint venture scheme is the lack of security of tenure. Does it follow then that such arrangements are against the interests of tenant farmers and the landlord and tenant system? The answer is—not necessarily at all. Indeed, these arrangements should, and undoubtedly will, develop to become an important part of a dynamic structure of land tenure. There are many tenants who can extend their businesses from the secure base of their tenancy and take on the exploitation of another land asset. It is a viable business proposition and there is no reason whatsoever, on those grounds, to oppose the principle. That should not, however, be taken to mean that unqualified approval should be given by the industry as a whole. The incentive for designing these schemes has been to circumvent the provisions of security of tenure laid out in current agricultural legislation. One can understand why, but that is not a good reason by itself. We have spent many years developing legislation with sound objectives in mind, and it follows that we should not go overboard in supporting alternatives just as a means of avoiding that legislation. They must be valid and stand up in their own right. Just as important, we should be asking whether the 1986 Act requires further modification to accommodate what is seen by many landowners as a fault in the system.

Grass-keep letting fulfils a traditional need and, provided the occupation sticks to grazing and mowing, poses no particular problems and serves the industry well. Short-term occupancy for arable cropping is equally common in the main arable and vegetable areas, and most people would say that, as with grass keeping, there is no problem. It is unfortunate that there is a degree of doubt surrounding this very valuable practice. The doubt can be avoided by ensuring that the user of the land does not have exclusive possession of the land; but, if you are not careful, that takes you pretty close to some form of joint venture. Alternatively, application may be made for a Ministry licence under the five-year scheme referred to earlier. However, the necessary form filling and the associated inevitable delay make this a clumsy solution to what is after all a very simple problem. The sensible thing to do would be to combine short-term arable cropping with grass-keep letting, but I must repeat, that is not the case at the present.

The better, and potentially much more important, use of the Ministry licensing system lies with the provision of short-term 'starter' tenancies. There is a real, and almost desperate, need for such tenancies to enable young entrants to gain the necessary experience which they can offer to get a full tenancy.

The Statutory Smallholdings Scheme operated by the County Councils was designed partly to fill this need but, for a variety of reasons, it has not worked quite like that. Also, particularly in the south-east commuter belt, there are areas of grassland lying around residential houses which are admirably suited for part-time farming, livestock grazing and fattening, or the keeping of horses. The grazing season let does not really fit the bill, and yet the owner does not want, for very good reasons, to be encumbered with a tenancy. There is a lot to be said for an extension of this system of Ministry five-year licensing, which amounts to independently controlled five-year term tenancies.

CHAPTER 15

The Future

I have spent quite a long time considering just how this book should be finished off. To a limited degree, it is a textbook with the objective of providing information, of preventing mistakes, and written therefore for the tenant, or tenant to be, in the hope that he will be that much better equipped to manage the part of his business concerned with the occupation of land belonging to his landlord. So in this concluding chapter, I could say—that's it, I have done my best to 'chart the seas' and it is now up to you and your professional advisers.

I don't think that that is quite good enough for, in another sense, it is not a textbook, but rather a commentary. Certainly I have done my utmost to ensure that it is factually correct. But it would be wrong to assume therefore that all the answers are within these covers. For that you have to go to Scammel and Densham, or Muir Watt, those admirable 'bibles' of landlord and tenant law. However, that is not the need I have identified, I hope correctly, which is that so many badly want what might be called a 'route map' rather than a bible. The textbooks are for the professional, be he lawyer or valuer.

So, if only partly a textbook, what lies in its other part? My belief is that, in a typically British way, we have allowed a system of land tenure to evolve over the centuries which has suited our people and our countryside extraordinarily well. It has provided the background against which our country houses, the great landed estates, the country market towns, the villages

89

and the individual farms have come together to form the rural scenery and its beauty which it is now becoming so fashionable to appreciate. What has not been always fully realised is how important the balance of land tenure as it has evolved is to the whole. Part of that balance is the unique division of responsibility which we call the landlord and tenant system. I am a fan of that system, but no fan should be so blind in his devotion that he cannot see the faults and the deficiencies; nor does it absolve him from the responsibility of trying to create something better.

So the question needs to be asked—is the system working? To the extent that, in comparison with other forms of land occupancy, it has been in decline, the answer must be in the negative. To the extent that an active search continues to find other forms of tenure, that too must be considered as evidence of deficiencies to be remedied. The reason is quite clear, and it lies in the way that the private letting landowner has been put at a disadvantage both as a result of an over-emphasis on the need for security of tenure, and as a result of penal taxation. What is far from clear, to me at least, is that we need to accept that as a consequence of these disadvantages, the days of the landlord and tenant system are numbered. Far better, and far more constructive, to continue our efforts at remedying the mistakes of the last forty years, a process which made a good start in the form of the 1984 and 1986 Agricultural Holdings Acts.

DOES IT MATTER?

There are those who say that the effort is not worthwhile: that the landlord and tenant system is a carry-over from a social order in the countryside which is no longer relevant; that it is based upon resident landowner and working farmer and, as such, is autocratic in its concept; that, moreover, it is insufficiently flexible to meet the needs of the capital-intensive business that modern farming has now become; and that the entrepreneurial flair of the modern businessman-farmer is better served by either owner occupation or short-term relationships with a landowner that is likely to be some kind of financial institution.

Quite apart from the dreadful image that this conjures up, I believe that this is rubbish, if only because it ignores the long-term nature of farming and land management, to say nothing of the intrinsic need for the maintenance of the quality of life in the countryside. The structure of land tenure does matter very much indeed and far beyond the relatively narrow, but nonetheless essential, bounds of financial return. The answer, it seems to me, is to maintain a balance; and there is no evidence to suggest that landlord and tenant are not, and cannot continue to be, an essential part of that balance.

WHAT SHOULD BE DONE?

The fundamental reason for the formation of the Tenant Farmers Association in 1981 was that there was no organisation which single mindedly looked after the interests of tenant farmers. The case for the tenant, and therefore by implication the case for the landlord and tenant system, was going by default. This can be taken as a criticism of the NFU, who saw that as their role. But my view is that it was always naive to expect that the NFU could fulfil this special role for tenants, any more than they could act for landowners in the way the CLA does. The NFU has a far wider duty to be responsible for the broad economic structure and prosperity of the industry as a whole.

My experience, serving as Vice Chairman and then Chairman of the TFA, bears out our original conviction that a close and continuing dialogue between the TFA and the CLA is absolutely vital. It is, of course, impossible to prove, but I have the feeling that some of the worst mistakes in the past could have been avoided had that community of interest taken physical shape much earlier. I am happy to say that successive presidents of the CLA with whom I worked shared this view and our determination to remedy past mistakes as well as to avoid future ones.

So my first answer to the question of what should be done is that every effort must be made to strengthen the links between the CLA and the TFA. That is not to say that the two organisations are always going to agree. Of course not, but they have far more in common than what divides them, and that is a solid base on which to build. To those of my fellow tenants who say that the TFA's prime duty is to pressurise rents downwards, I would reply that that is a misconception. The Association's role must be to enable its members to negotiate a fair rent. Rents which are too low, just as much as rents which are too high, damage the mutual partnership that is at the core of a successful landlord and tenant relationship. You cannot have tenants without landlords, and vice versa. Greed is self defeating, whoever exercises it.

IMPROVING THE TERMS OF TENANCY

So what should the CLA and the TFA be talking about? Clearly, all those problems which come up day by day in agricultural politics. How should Government grants for tree planting be fitted into the tenancy? Can set-aside, voluntary or obligatory, be shared between landlord and tenant, and what effect should this have on the rent? And, if faced with new quota proposals, what action can be jointly taken to minimise the damage and treat both parties fairly and justly?

Important as these day to day matters are—and they are vitally important—we should not be distracted by the need to survive today from constructive planning for the morrow. Fundamental to the average landowner's reluctance to let a farm has been the financial sacrifice inherent in him losing access to the vacant possession value. He does, after all, own the land and there is no law in a free country which says that he must let it to a tenant. If that sacrifice is too great in relation to the benefit that he gets from a risk-free income in the form of rent, then he will seek other ways of using his land, be they taking the land in hand himself or as one of the various forms of joint venture. And that financial argument is further strengthened by fears that noises from the left of British politics could lead one day to a return to policies designed to penalise him.

TERM TENANCIES

These fears and arguments lead some to propose that the industry should pursue the introduction of term tenancies, that is to say, tenure for a pre-determined number of years, be it 10 years or 15 or maybe 21, after which vacant possession is given. It is said, I believe quite rightly, that this would lead to far more farms coming on the market to let, and it would consequently lead to a resurgence in the landlord and tenant system.

The TFA has consistently argued against this, and with justification, for a number of reasons. It would strike at the root of security of tenure and the tenant's long-term interest in developing the farm by farming it as well for the future as for today. Furthermore, it would also destroy the standard tenancy for no landowner, or his agent, would consider letting on a normal tenancy if term tenancies were available. We should think very carefully indeed, and for a very long time, before we consider putting the clock back as far as that.

There is, furthermore, a more practical reason for opposing term tenancies. That is the risk, which is virtually a certainty, that before long someone, somewhere, would lose possession of his farm and have nowhere to go. No matter that this would be perfectly legal or that he would only have himself to blame for not anticipating the end of his term. It is not difficult to imagine the press publicity surrounding such a case—'Innocent family turned out in the rain by wicked landlord'—and the danger that that would lead inevitably to a return to 1976-style protection of tenants. Anyone who has the long-term interests of landlord and tenant at heart should shudder at the prospect, and it ought to be enough to turn anyone away from term tenancies.

RETIREMENT TENANCIES

We must not, however, be purely negative in opposing term tenancies, because the underlying reason for proposing them remains valid. The owner of agricultural land has a case when he says that he must be sure of getting vacant possession at some future and definite date, and that it is unreasonable, and indeed counter-productive, to give the tenant too much security, too much shelter from economic reality.

It is the view of the TFA that this problem can be solved by the introduction of retirement tenancies. Briefly, the proposal is that future tenancies should include a clause granting the landlord the right to serve an Incontestible Notice to Quit when the tenant reaches the age of 65. There is a clear precedent for this in the County Council Smallholdings clause. It is, of course, a form of term tenancy in that it has a definitive end—the owner knows exactly when he will get possession and can plan his estate finances accordingly. But it is a term tenancy without the political risk attendant on a tenant losing possession during the course of his working life. Moreover, provided the terms of the tenancy are linked to adequate provisions for funding retire- ment—and that is an essential proviso—such a tenancy would be very much in line with current social thinking. In con- sequence, I am advised that it would be relatively easy to get such an amendment to legislation through both Houses of Parliament. I am convinced that retirement tenancies would lead to many more farms being let, particularly as many owners now realise the difficulties of farm management on in-hand estates that have grown too big and unwieldy.

THE OWNERSHIP OF LAND

In a book which is primarily written from the point of view of a tenant, it may seem inappropriate to include a section on land ownership. But a tenant surely has a basic reason for being interested in who owns the land. Anyone reading this chapter, or indeed the whole book, cannot fail to have noticed that I am an advocate of balance and an absence of extremes. That too applies to land ownership. Since the days of the enclosures, we have had such a balance: institutional ownership in the form of the Church, the Crown and the Oxford and Cambridge colleges; the landed estates, both large and small; and, at the other end of the scale, the individual owning his own farm. It is only relatively recently that City investment money has come into land, mainly in the form of pension funds seeking a safe refuge from inflation and looking for capital appreciation. Some of it went into managed farming operations, often hand in hand with large management companies, whilst the majority went into the purchase of farms already tenanted, and often bought on a sale and lease back arrangement.

Does it really matter who owns the land so long as a structure exists to exercise some control on its management? That structure exists in the form of the Agricultural Holdings Acts, so have we any cause for concern at the changing patterns of land ownership? I believe that it does matter, and that there is some cause for concern. To explain that I have to go back to what I call the organic whole of the countryside—the inter-relationship of church and school, of parson and headmaster; of landowner, occupier and tenant; of farming and other rural industries; of farmer, farm worker and all the other residents who live in a village. How this inter-relationship functions affects the health of the countryside. More important still is my belief that the health of the countryside is as essential to the well-being of the nation as the state of stock exchange—and some would say, more so. So I feel instinctively that it is a pity if we see a substantial involvement in land ownership by those whose interests lie elsewhere, and who do not choose to live in, or involve themselves with, the life of the countryside. Happily, it seems as if this problem will find its own solution, as much of the City money that came into land ownership is not earning the return which over-optimistic salesmen predicted—something we could have told them before they started!

It would be wrong, however, to leave the subject on that emotional, and rather smug, note, for we should be asking ourselves a much more constructive question. How can we encourage the right type of caring and resident landowner? The answer must surely lie in continuing efforts to reverse the trends in penal taxation that have struck at the roots of private land ownership—another joint task for the CLA and the TFA.

CONCLUSION

Let me finally restate my conviction that in the landlord and tenant system we have a unique system of land tenure, which we have treated with a cavalier disregard for its health. But it is far from lost: indeed the reverse is true, and I believe that we shall see a return to the situation where around 40 per cent of our land is held in tenancy.

I can express that belief whilst at the same time stating my further conviction that the relationship is no longer, in any sense, feudal; nor is it a relationship between two unequal parties; it is one which is regulated by law which is often complex and which is fraught with pitfalls. Understanding the essentials of that law is as much part of the business of farming as is stock sense or green fingers.

Appendix I

Bibliography

Joint Ventures in Farming; report by Richard Stratton, from CLA Publications, 16 Belgrave Square, London

Share Farming, by Richard Stratton, Michael Gregory and Richard Williams, from CLA Publications, 16 Belgrave Square, London

Agricultural Holdings Act 1986, an annotated edition by James Muir Watt, published by Sweet and Maxwell, London

Agricultural Holdings Act 1984: The Practitioner's Companion, by Donald Troup, published by Surveyors Publications, 12 Great George Street, London

Muir Watt's Agricultural Holdings, by James Muir Watt, published by Sweet and Maxwell, London

Scammell and Densham's Law of Agricultural Holdings, by H. A. C. Densham, published by Butterworth Press

APPENDIX 2

Agricultural Land Tribunals

Agricultural Land Tribunals (ALT) were first established under the Agriculture Act 1947. Amendments have been made since then, but their purpose remains the same; that is, to deal with what are known as the 'principal rules' under the Agricultural Land Tribunals (Rules) Order of 1978, and with succession provisions under a similar Order of 1984.

Reference has been made to the use of the ALT in the text on several occasions. The most important of these are concerned with Notices to Quit and to Remedy, with improvements and related compensation, and with cases of disputed succession. As always, it is essential to seek skilled professional advice and to conform to the necessary procedures regarding time limits and the use of the forms of application appropriate to each case.

England and Wales are divided into eight areas, each with its own ALT. The chairman is appointed by the Lord Chancellor and must be a barrister or solicitor of not less than seven years' standing. The Lord Chancellor is also required to draw up a panel of similarly qualified persons who may act as deputy chairmen, as well as two panels of persons appearing to him to represent farmers and landowners respectively, these latter panels being formed from nominations submitted by the CLA and the NFU. Each hearing has a specially constituted Tribunal consisting of the chairman and one person nominated by the chairman from each of the landowners' and farmers' panels.

The ALT has powers to call evidence and cross-examine witnesses on oath, and the provisions of the County Court rules apply in the same way as for arbitrations. The decision of the ALT, which may be a majority decision, must be given in writing together with a statement or reasons.

99

APPENDIX 3

Calculation of Productive and Related Earning Capacity

Chapter 6 deals with rent review procedure. One of the planks on which that review is built is the calculation of the productive capacity of the farm and its related earning capacity. As explained in that chapter, this is a new element in the negotiations leading up to a possible arbitration. Many of those who will be involved have much to learn about a proper and simple method of preparation and presentation. In particular, I am concerned that tenants should be able to provide their valuers with the necessary accurate information on which this part of the case can be built. I have thought it right therefore to set out the bare bones of such a method for two hypothetical farms, one arable and the other dairy. In doing so, I have been very aware of the danger of writing another book. I have thus kept the approach extremely simple, some may say too simple, in order to illustrate the essential framework. Each individual case is, however, complex to some degree or another; and not only complex, but different in matters of detail. Once more the skill of the valuer comes into play, but his expertise will be that much more difficult to exercise if he is not provided with the fullest possible information.

To put flesh on the bones of the two examples, I have had to include figures for yield, and for variable and fixed costs. These are taken from the *John Nix Farm Management Pocket Book 1987* and the *ABC Costing Book No. 25 1987*.

EXAMPLE 1
Slumberdown Farm: Rent Review 1988
Total acres: 520   Croppable acres: 500   Soil type: chalk

The actual and historic productive capacity of the farm is shown in Table 1.

101

Table 1   Actual yields

Crop	Acres	Total Tonnes	Yield/acre	Total cash	Cash/ acre
Wheat					
1983	200	578	2.89	£70516	£352
1984	200	760	3.80	£88920	£444
1985	200	550	2.75	£82368	£411
1986	200	624	3.12	£81120	£405
1987	200	486	2.43	£54432	£272
	Average	599	2.99	*£75471*	*£376*
Barley					
1983	100	263	2.63	£29719	£297
1984	100	342	3.42	£37278	£372
1985	100	242	2.42	£32967	£329
1986	100	297	2.97	£33264	£332
1987	100	212	2.12	£23108	£231
	Average	271	2.71	*£31267*	*£312*
Beans					
1983	—				
1984	—				
1985	—				
1986	50	87	1.74	£16095	£322
1987	50	105	2.10	£19635	£392
	Average	96	1.92	*£17865*	*£357*
Sugar Beet					
1983	100	1732	17.32	£46417	£464
1984	100	2143	21.43	£57861	£578
1985	100	1589	15.89	£42585	£425
1986	100	1838	18.38	£49993	£499
1987	100	1841	18.41	£49504	£495
	Average	1828	18.28	*£49272*	*£492*
Oilseed Rape					
1983	100	105	1.05	£33600	£336
1984	100	165	1.65	£46200	£462
1985	100	148	1.48	£43032	£430
1986	50	75.5	1.51	£21668	£433
1987	50	79.5	1.59	£17887	£357
	Average	114.6	1.45	*£32477*	*£403*

Appendix 3

Note the two cash columns set in bold type. This is part of the information which you, the tenant, should have in order to evaluate your present and future rent. It is not, however, relevant information for the review for two reasons. One, because it is based on historical prices; and two, because it reflects the tenant's own ability to sell. Remember always that the review must be determined on the basis of the farm being farmed by the hypothetical 'competent' tenant. The physical yield information is, however, relevant in that it gives the best indication of the farm's productive capacity.

How does the capacity of Slumberdown Farm compare with similar farms in the area?

Table 2   Comparable yields on farms in the area (tonnes/acre)

Crop	Year	Slumberdown	Farm A	Farm B	Farm C	Overall Average
Wheat	1984	3.80	3.60	3.90	2.95	
	1985	2.75	3.00	3.20	2.68	
	1986	3.12	3.30	3.34	2.81	
	Average	*3.22*	*3.30*	*3.30*	*2.81*	*3.16*
Barley	1984	3.42	3.10	2.80	2.45	
	1985	2.42	2.80	2.62	2.53	
	1986	2.97	3.20	2.84	2.73	
	Average	*2.93*	*3.00*	*2.75*	*2.57*	*2.81*
Sugar Beet	1984	21.43	23.40	—	18.62	
	1985	15.89	16.32	—	14.97	
	1986	18.38	19.41	—	16.32	
	Average	*18.57*	*19.71*	*—*	*16.64*	*18.31*
Oilseed Rape	1984	1.65	1.58	—	1.32	
	1985	1.48	1.51	—	1.41	
	1986	1.51	1.49	—	1.25	
	Average	*1.55*	*1.53*	*—*	*1.33*	*1.47*

Notes:
All above weights in tonnes actually sold per OS acre.
Farm A — 872 acres
Farm B — 297 acres
Farm C — 620 acres
Clearly, all four farms are on very good land and are well farmed.

We must now evaluate Slumberdown Farm as it might be farmed in the hands of an average 'competent' tenant practising a system suitable for the holding. For physical yield, I have used the average yields shown in Table 2.

Table 3   Calculation of gross margin

1987 Cropping	Acres	Output	Average Variable Costs	Gross Margin
Wheat	200	3.16 T × £110 = £347 × £69400	£87 = £17400	£260 = £52000
Barley	100	2.91 T × £105 = £305 × £30500	£69 = £6900	£236 = £23600
Beans	50	1.92 T × £170 = £326 = £16300	£51 = £2550	£275 = £13750
Sugar Beet	100	18.31 T × £27 = £404 = £49400	£162 = £16200	£332 = £33200
Oilseed Rape	50	1.47 T × £225 = £330 = £16500	£100 = £5000	£230 = £11500
	500 per acre	£182100 *£364*	£48050 *£96*	£134050 *£268*

From this gross margin calculation, we now need to deduct the fixed costs of the enterprise in order to arrive at the figure of the 'cake', that is to say, the sum which is available to be divided between landlord and tenant. Whilst it is correct to use independent averages for variable costs, I believe that with fixed costs it is better to use the farm's actual figures. This is quite simply because they can vary so widely that any averages tend to be meaningless. So, more analysis of the farm's accounts is necessary (how much easier if it is done every year as a matter of course by your accountant).

Let us imagine that, in the case of Slumberdown Farm, the total fixed costs amount to £107,500 or £215 per acre. The tenancy is based on the Model Clauses, and the rent included in the figure is £50 per acre. The £215 does, however, include certain items which must be taken out: e.g. the tenant's total drawings must be reduced to the level of his actual work on the farm, say, £10,000 p.a.; and, the total cost of borrowings. It must be said there is some dispute as to how much of this is allowable. My own view is that interest on overdraft for non-hard-core borrowing is a legitimate expense of the business and should remain.

After these adjustments have been made, the allowable fixed costs are down to £97,760 or £195 per acre.

CALCULATION OF THE CAKE

The 'cake' can be calculated as follows:

the gross margin per acre (£268) minus the fixed costs less rent, i.e. £268 − (£195 − £50) = *£123 per acre*

So, on the basis of our calculations so far, we have a sum to divide between the parties. But wait a minute! Let us not forget that the rent is being settled for the *next three years.* Our output figures have used actual 1987 prices. How do we, and, more important, how does the arbitrator judge the prospects for the period to 1991? A difficult question indeed! And one which many are tempted to put to one side as pure speculation. This is hardly being fair and reasonable, however, as, in the past, when farm profits were expected to rise, negotiations always, and properly, anticipated that rise. The 'prudent' tenant in 1988 can only anticipate falling returns, and he will no doubt make the point with some vigour. But falling to what level? Each individual must make his own calculation, aided as far as possible by economic predictions. In doing so, it is well to remember that relatively small reductions in end product prices can have a greatly disproportionate effect on the available cake.

So we could perhaps have three different cakes:

- the 1988 cake of £123 per acre
- an arguable cake based on the supposition that the cake will fall in value by 7 per cent per annum:
    1989 = £114
    1990 = £106
    1991 = £98
    average = £106 per acre.
- another arguable cake based on the supposition that end product prices will fall by 5 per cent per annum, but all costs remain static.

105

Using the previous figures, output will fall from a total of £182,100 to:

1989 £172995 and the gross margin to £250 per acre
1990 £164345 and the gross margin to £232 per acre
1991 £156127 and the gross margin to £216 per acre

We then deduct fixed costs less rent, as before, i.e.

1989 £250 − £145 = £105
1990 £232 − £145 = £87
1991 £216 − £145 = £71

Thus the average of these is a cake of £87.

CUTTING THE CAKE

How should it be cut? There is an assumption in many circles that a 50/50 split is reasonable, but it should be emphasised that there is nothing sacred about such an arbitrary figure. It could be said, with some considerable logic, that, as the cake gets smaller, the tenant will have to have a larger slice if he is to survive.

But let us assume, for the sake of this example, that the 50/50 split is agreed. Then,

£123 ÷ 2 = £61.50 per acre
£106 ÷ 2 = £53 per acre
£ 87 ÷ 2 = £43.50 per acre

TENANT'S IMPROVEMENTS

Still we have not finished, as we have to place a value on the tenant's improvements. Here we face another area of lack of precision, for there is no prescribed method of how to determine an increase in rental value resulting from such an improvement. Different valuers have different methods. There is, for instance, the so-called 'black patch' approach, where the improvement is ignored, and an attempt is made to estimate the value of the output without the benefit of the improvement. Much too difficult in my view.

Alternatively, there is what is known as the 'annual equivalent'. Imagine that, seven years previously, the tenant paid for a new

106

building: a substantial portal framed structure, with concrete floor and block walls, at a gross cost before grant of £27,000. It was erected with the landlord's written consent, although this is not an essential condition for rent calculation as it is for end of tenancy compensation. It has an estimated life of 25 years, of which 18 still remain.

There are two methods of calculating the annual equivalent:

Firstly, on replacement value. The 1988 cost of erecting an identical building can be determined by quotation, but could easily be £38,000.

thus: 18/25 × £38000 = £27360
    at a nominal interest charge of say 10%   = £2736 p.a.

or secondly, on an annual interest charge of say 10% on the original price

thus: 10% of £27000   = £2700 p.a.

We can now deduct this from the previously calculated 'rental values':

1 the '1987' value of £61.50 per acre × 500 acres   = £30750
  less, say, £2700 = £28050 on 500 acres   = *£56.10 per acre*

2 the first 1989/91 value of £53 per acre × 500 acres = £26500
  less the £2700 = £23800 on 500 acres   = *£47.60 per acre*

3 the second 1989/91 value of £43.50 per acre × 500 acres
                                                    = £21750
  less the £2700 = £19050 on 500 acres   = *£38.10 per acre*

So we have finally arrived at a rent calculated on the basis of the productive and related earning capacity. Or rather three different rents depending on an estimation of future prospects. Fertile ground for negotiation taking over from arithmetic!

In any enthusiasm that we may have for arithmetic, don't let us forget that the arbitrator is also instructed to take account of other factors (see Chapter 6) of which the rent actually paid on comparable farms is of the greatest importance.

Example 2
Mill Farm: Rent Review 1988
Total acres: 100   Specialist dairy farm   All permanent pasture

A similar exercise can be carried out on any farm, but let us see how the final calculations apply on this specialist dairy farm. Note that I am not listing here the records of previous production, nor any details of comparable farms; but it should be realised that this is every bit as important here as it was with Slumberdown Farm.

When we come to the evaluation of Mill Farm, the first hurdle is that of milk quota. Clearly, the earning capacity of the farm, as a dairy farm, is limited first and foremost by quota. It is fundamental therefore to establish right at the outset what level of quota is *allocated* to the farm in question. This is not necessarily the same quantity as the quota registered on the holding. If some part of the milk production is being supported on land other than that on the subject holding, then that part of the production must be discounted. If, for this reason, there is insufficient quota for the holding to be evaluated totally as a dairy farm, then an imaginary secondary enterprise must be included to take the place of the 'lost' litres, e.g. beef cattle. There may also be the other circumstance where the tenant has bought or leased additional quota in order to 'top up' the allocation. Clearly, these additional litres must be similarly disregarded.

In this example however it is assumed that there is sufficient quota, and we can look at two different levels of production, using average costings:

80 cows selling 5200 litres @ 16.6 p	£863 per cow
less variables including forage	£362 per cow
	£501 per cow
20 heifers up to calving less calf value	£475
less variables including forage	£250
	£225 per heifer unit
Total Gross Margin £501 × 80	£40080
£225 × 20	£ 4500
	£44580 = £445/acre

FIXED COSTS

As with Slumberdown Farm, it is preferable to list the farm's actual fixed costs. But with a small dairy farm run with family labour, it is essential that *all* labour costs are included. The category of 'Specialist Dairy Farms up to 125 acres' (John Nix) shows average fixed costs of £342 per acre including £57 per acre rent. This level of fixed costs is by no means unusual on this size and type of farm, and that is what I shall use.

Thus, the calculation of the cake:

£445 − (£342 − £57) =	£160 per acre
on a 50/50 split =	£ *80 per acre*

But it should be said right away that for 100 acres to carry 80 cows and rearing 20 heifers to down calving is a stocking rate and performance way above the average of the ordinary 'competent' tenant.

If the 100 acres were only supporting 60 cows and 20 heifer replacements, and the cows were averaging 4900 litres per cow, then:

60 cows selling 4900 litres @ 16.6 p	£813 per cow
less variables including forage and less concs.	£332 per cow
	£481 per cow
20 heifers as before	*£225 per heifer unit*
Total gross margin £481 × 60	£28860
£225 × 20	£ 4500
	£33360 = £333/acre

And taking fixed costs as before, the calculation of the cake:

£333 − (£342 − £57) =	£48 per acre
on a 50/50 split =	*£24 per acre*

Remember that after these calculations we still have to deduct the value of any improvements done by the tenant.

These figures illustrate just how dependent the small dairy farm is on high performance per cow and per acre and also on the use of family labour. The 'John Nix' figures used for fixed costs of £342 per acre include £146 per acre for paid labour, which amounts to £14,600 for Mill Farm. If this figure is not included in the actual fixed costs for the farm, then a completely false picture will emerge.

Finally, as with Slumberdown Farm, we have to consider future prospects. What will the level of quota be, and what will be the price of milk, in three years' time?

The evaluation of rent payable on a specialist dairy farm is full of endless complications. The examples given are, of course, an over-simplification; but they do underline the fundamental necessity of accurate record keeping so as to enable the valuer to be fully equipped for his job.

Index

Index

Landlord and tenant system (*cont.*)
  maintaining balance, 89–90
  prosperity, 2
Law, development, 11–14
Lease
  definition, 19
  differences from tenancy, 20–22
  reverts to standard tenancy, 19–20
  or tenancy agreement, 17–22
  term, 19
Licences, Ministry, 84, 87–88
Lincolnshire Agreement or Custom, 7, 17
  and tenant's improvements, 51

Manorial system, 5–6
Marriage value, 26, 30
Medieval farming systems, 5
Milk quotas, 77–81
  and *Agriculture Act 1986*, 78–79
  attachment, 78
  compensation basis, 79–81
  complications, 77–81
  disposal, 77
  excess, 80
  maintaining, 77
  standard, 79–80
Ministry licences, 84
Model Clause agreement for repairs and
  insurance, 57

National Farmers Union, aims, 13
Negotiations for rent determination, 31–
  32
Notices, 48
  to quit, 67–72
    circumstances, 68–69
    compensation, 70–72
    contestable, 70
    incontestable, 68–69
  to remedy, 59–60
  serving, rents and leases, 21–22
  under 1986 Agricultural Holdings Act,
    48

Obligations, 55–60
Open market value rent, 21
Ownership of land, 95–96

Partnerships and share farming, 84–85
  advantages, 86–88

Productive capacity and related earning
  capacity, appendix 3, 101
  and rents, 28–30

Quota complications, 77–81

Records, essential, 73–76
Rents
  definitions, 20–22
    fixed, 20
    open market, 21
    serving of notices, 21–22
    upwards only, 20–21
  determination, and arbitration, 23–37
    Agricultural Holdings Act 1986, 23
    arbitration, 34–35
    arbitrator, appointment, 32
    award, 35–36
    character and situation of holding,
      27–28
    comparable holdings, 30–31
    costs, 36–37
    negotiations, 31–32
    productive capacity and related
      earning capacity, 28–30
    section 12 notice, 24–25
    statement of case, 33–34
    terms of the tenancy, 26–27
  reviews, keeping records, 74–75
  succession, 48
Repairs, responsibilities, 55–56
Residence in the farm house,
  responsibilities, 58
Responsibilities of tenant, 55–60
  notices to remedy, 59–60
Retirement
  and succession, 46–47
  tenancies, 94–95
Roofs, landlord responsibilities, 56
Rules of Good Husbandry, 12, 71

Section 12 notice, 24–25
Share farming, 84–85
Short-term tenancies, 65
Soil testing before renting, 63
Statutory Smallholdings Scheme, 87–88
Stocking responsibilities, 55–56
Stone walls and banks, responsibilities,
  58
Structures, landlord responsibilities, 56

112

Index

FARMING PRESS BOOKS

Below is a sample of the wide range of agricultural and veterinary books published by Farming Press. For more information or for a free illustrated book list please contact:

Books Department, Farming Press Ltd, Wharfedale Road, Ipswich IP1 4LG, Suffolk, Great Britain.

A Way of Life: Sheepdog Training, Handling and Trialling
H. Glyn Jones and Barbara Collins

A complete guide to sheepdog work. The story of Glyn Jones' life with sheepdogs is presented as an integral part of his practical advice on training, handling, trialling and breeding.

Intensive Sheep Management
Henry Fell

An instructive, practical volume giving the best of traditional practices along with new management, feeding and housing techniques.

The Growing and Finishing Pig: Improving its Efficiency
P. R. English, S. H. Baxter,
V. R. Fowler, W. J. Smith

A large, comprehensive volume which explores in detail the factors that control the efficiency of the pig from weaning to slaughter.

Outbursts
Oliver Walston

A provocative series of articles by one of Britain's best-known arable farmers commenting on the exceptional conditions that characterised the period starting with the Great Drought of 1976.

Udderwise
Emil van Beest

A collection of cartoons presenting the lighter side of bovine performance.

Farm Woodland Management
John Blyth, Julian Evans, William E. S. Mutch, Caroline Sidwell

A timely compendium for farmers in which all aspects of trees on the farm are considered.

Outdoor Pig Production
Keith Thornton

The first practical handbook on modern techniques of outdoor pig keeping.

A Veterinary Book for Dairy Farmers
Roger Blowey

A new, enlarged edition of this very useful book which emphasises preventive medicine.

Farm Building Construction
Maurice Barnes and Clive Mander

Practical information covering all aspects of farm building, from initial planning onwards.

Crop Nutrition and Fertiliser Use
John Archer

Gives details of each nutrient and then specific requirements of temperate crops, ranging from grassland and cereals to vegetables and fruit stock. New edition.

Farming Press also publish three monthly magazines: *Dairy Farmer, Pig Farming* and *Arable Farming*. For a specimen copy of any of these magazines please contact Farming Press at the address above.